Lincoln
Gold Portfolio
1949-1960

Compiled by
R.M. Clarke

ISBN 1 85520 0163

Distributed by
Brooklands Book Distribution Ltd.
'Holmerise', Seven Hills Road,
Cobham, Surrey, England

Printed in Hong Kong

BROOKLANDS BOOKS

BROOKLANDS BOOK SERIES
AC Ace & Aceca 1953-1983
Alfa Romeo Alfasud 1972-1984
Alfa Romeo Alfetta Coupes GT GTV GTV6 1974-1987
Alfa Romeo Guilia Berlinas 1962-1976
Alfa Romeo Giulia Coupes 1963-1976
Alfa Romeo Spider 1966-1987
Allard Gold Portfolio 1937-1958
Alvis Gold Portfolio 1919-1969
American Motors Muscle Cars 1966-1970
Aston Martin Gold Portfolio 1972-1985
Austin Seven 1922-1982
Austin A30 & A35 1951-1962
Austin Healey 3000 1959-1967
Austin Healey 100 & 3000 Col No.1
Austin Healey 'Frogeye' Sprite Col No.1 1958-1961
Austin Healey Sprite 1958-1971
Avanti 1962-1983
BMW Six Cylinder Coupes 1969-1975
BMW 1600 Col. 1 1966-1981
BMW 2002 1968-1976
Bristol Cars Gold Portfolio 1946-1985
Buick Automobiles 1947-1960
Buick Muscle Cars 1965-1970
Buick Riviera 1963-1978
Cadillac Automobiles 1949-1959
Cadillac Automobiles 1960-1969
Cadillac Eldorado 1967-1978
High Performance Capris Gold Portfolio 1969-1987
Chevrolet Camaro & Z-28 1973-1981
High Performance Camaros 1982-1988
Chevrolet Camaro Col No.1 1967-1973
Camaro Muscle Cars 1966-1972
Chevrolet 1955-1957
Chevrolet Impala & SS 1958-1971
Chevrolet Muscle Cars 1966-1971
Chevelle and SS 1964-1972
Chevy EL Camino & SS 1959-1987
Chevy II Nova & SS 1962-1973
Chrysler 300 1955-1970
Citroen Traction Avant 1934-1957
Citroen DS & ID 1955-1975
Citroen 2CV 1949-1988
Cobras & Replicas 1962-1983
Corvair 1959-1968
High Performance Corvettes 1983-1989
Datsun 240Z 1970-1973
Datsun 280Z & ZX 1975-1983
De Tomaso Collection No.1 1962-1981
Dodge Charger 1966-1974
Dodge Muscle Cars 1967-1970
Excalibur Collection No.1 1952-1981
Ferrari Cars 1946-1956
Ferrari Cars 1973-1977
Ferrari Dino 1965-1974
Ferrari Dino 308 1974-1979
Ferrari 308 & Mondial 1980-1984
Ferarri Collection No.1 1960-1970
Fiat-Bertone X1/9 1973-1988
Fiat Pininfarina 124 + 2000 Spider 1968-1985
Ford Automobiles 1949-1959
Ford GT40 Gold Portfolio 1964-1987
Ford Fairlane 1955-1970
Ford Falcon 1960-1970
High Performance Mustangs 1982-1988
Ford Cortina 1600E & GT 1967-1970
Ford RS Escorts 1968-1980
High Performance Escorts Mk1 1968-1974
High Performance Escorts Mk II 1975-1980
Honda CRX 1983-1987
Hudson & Railton 1936-1940
Jaguar Cars 1957-1961
Jaguar Cars 1961-1964
Jaguar Mk2 1959-1969
Jaguar E-Type Gold Portfolio 1961-1971
Jaguar E-Type 1966-1971
Jaguar E-Type V-12 1971-1975
Jaguar XKE Collection No.1 1961-1974
Jaguar XJ6 1968-1972
Jaguar XJ6 Series II 1973-1979
Jaguar XJ6 & XJ12 Series III 1979-1985
Jaguar XJ12 1972-1980
Jaguar XJS Gold Portfolio 1975-1990
Jaguar XK120.XK140.XK150 Gold Portfolio 1948-1960
Jensen Cars 1946-1967
Jensen Cars 1967-1979
Jensen Healey 1972-1976
Jensen Interceptor Gold Portfolio 1966-1986
Lamborghini Cars 1964-1970
Lamborghini Cars 1970-1975
Lamborghini Countach Col No.1 1971-1982
Lamborghini Countach & Urraco 1974-1980
Lamborghini Countach & Jalpa 1980-1985
Lancia Stratos 1972-1985
Land Rover 1948-1973 - A Collection
Land Rover Series II & IIa 1958-1971
Land Rover Series III 1971-1985
Land Rover 90 & 110 1983-1989
Lincoln Gold Portfolio 1949-1960
Lincoln Continental 1961-1969
Lotus and Caterham Seven Gold Portfolio 1957-1989
Lotus Elan Gold Portfolio 1962-1974
Lotus Elan Collection No.2 1963-1972
Lotus Elite 1957-1964
Lotus Elite & Eclat 1974-1981
Lotus Turbo Esprit 1980-1986
Lotus Europa 1966-1975
Lotus Europa Collection No.1 1966-1974
Lotus Seven Collection No.1 1957-1982
Marcos Cars 1960-1988
Maserati 1965-1970
Maserati 1970-1975
Mazda RX-7 Collection No.1 1978-1981
Mercedes 190 & 300SL 1954-1963
Mercedes 230/250/280SL 1963-1971
Mercedes Benz SLs & SLCs Gold Portfolio 1971-1989
Mercedes Benz Cars 1949-1954
Mercedes Benz Cars 1954-1957
Mercedes Benz Cars 1957-1961
Mercedes Benz Competition Cars 1950-1957
Mercury Muscle 1966-1971
Metropolitan 1954-1962
MG TC 1945-1949
MG TD 1949-1953
MG TF 1953-1955
MG Cars 1959-1962
MGA Roadsters 1955-1962
MGA Collection No.1 1955-1982
MGB Roadsters 1962-1980
MGB GT 1965-1980
MG Midget 1961-1980
Mini Moke 1964-1989
Mini Muscle Cars 1961-1979
Mopar Muscle Cars 1964-1967
Mopar Muscle Cars 1968-1971
Morgan Three-Wheeler Gold Portfolio 1910-1952
Morgan Cars 1960-1970
Morgan Cars Gold Portfolio 1968-1989
Morris Minor Collection No.1
Mustang Muscle Cars 1967-1971
Oldsmobile Automobiles 1955-1963
Old's Cutlass & 4-4-2 1964-1972
Oldsmobile Muscle Cars 1964-1971
Oldsmobile Toronado 1966-1978
Opel GT 1968-1973
Packard Gold Portfolio 1946-1958
Pantera Gold Portfolio 1970-1989
Plymouth Barracuda 1964-1974
Plymouth Muscle Cars 1966-1971
Pontiac Tempest & GTO 1961-1965
Pontiac GTO 1964-1970
Pontiac Firebird 1967-1973
Pontiac Firebird and Trans-Am 1973-1981
High Performance Firebirds 1982-1988
Pontiac Fiero 1984-1988
Pontiac Muscle Cars 1966-1972
Porsche 356 1952-1965
Porsche Cars in the 60's
Porsche Cars 1960-1964
Porsche Cars 1964-1968
Porsche Cars 1968-1972
Porsche Cars 1972-1975
Porsche Turbo Collection No.1 1975-1980
Porsche 911 1965-1969
Porsche 911 1970-1972
Porsche 911 1973-1977
Porsche 911 Carrera 1973-1977
Porsche 911 Turbo 1975-1984
Porsche 911 SC 1978-1983
Porsche 914 Gold Portfolio 1969-1976
Porsche 914 Collection No.1 1969-1983
Porsche 924 Gold Portfolio 1975-1988
Porsche 928 1977-1989
Porsche 944 1981-1985
Range Rover Gold Portfolio 1970-1988
Reliant Scimitar 1964-1986
Riley 11/2 & 21/2 Litre Gold Portfolio 1945-1955
Rolls Royce Silver Cloud 1955-1965
Rolls Royce Silver Shadow 1965-1981
Rover P4 1949-1959
Rover P4 1955-1964
Rover 3 & 3.5 Litre 1958-1973
Rover 2000 + 2200 1963-1977
Rover 3500 1968-1977
Rover 3500 & Vitesse 1976-1986
Saab Sonett Collection No.1 1966-1974
Saab Turbo 1976-1983
Shelby Mustang Muscle Cars 1965-1970
Stubebaker Gold Portfolio 1947-1966
Stubebaker Hawks & Larks 1956-1963
Sunbeam Tiger & Alpine Gold Portfolio 1959-1967
Thunderbird 1955-1957
Thunderbird 1958-1963
Thunderbird 1964-1976
Toyota MR2 1984-1989
Triumph 2000. 2.5. 2500 1963-1977
Triumph GT6 1966-1974
Triumph Spitfire 1962-1980
Triumph Spitfire Col No.1 1962-1980
Triumph Stag 1970-1980
Triumph Stag Collection No.1 1970-1984
Triumph TR2 & TR3 1952-60
Triumph TR4-TR5-TR250 1961-1968
Triumph TR6 1969-1976
Triumph TR6 Collection No.1 1969-1983
Triumph TR7 & TR8 1975-1982
Triumph Vitesse & Herald 1959-1971
TVR Gold Portfolio 1959-1988
Volkswagen Cars 1936-1956
VW Beetle Collection No.1 1970-1982
VW Golf GTi 1976-1986
VW Karmann Ghia 1955-1982
VW Kubelwagen 1940-1975
VW Scirocco 1974-1981
VW Bus. Camper. Van 1954-1967
VW Bus. Camper. Van 1968-1979
VW Bus. Camper. Van 1979-1989
Volvo 120 1956-1970
Volvo 1800 1960-1973

BROOKLAND ROAD & TRACK SERIES
Road & Track on Alfa Romeo 1949-1963
Road & Track on Alfa Romeo 1964-1970
Road & Track on Alfa Romeo 1971-1976
Road & Track on Alfa Romeo 1977-1989
Road & Track on Aston Martin 1962-1984
Road & Track on Auburn Cord and Duesenburg 1952-1984
Road & Track on Audi & Auto Union 1952-1980
Road & Track on Audi 1980-1986
Road & Track on Austin Healey 1953-1970
Road & Track on BMW Cars 1966-1974
Road & Track on BMW Cars 1975-1978
Road & Track on BMW Cars 1979-1983
Road & Track on Cobra, Shelby & GT40 1962-1983
Road & Track on Corvette 1953-1967
Road & Track on Corvette 1968-1982
Road & Track on Corvette 1982-1986
Road & Track on Datsun Z 1970-1983
Road & Track on Ferrari 1950-1968
Road & Track on Ferrari 1968-1974
Road & Track on Ferrari 1975-1981
Road & Track on Ferrari 1981-1984
Road & Track on Fiat Sports Cars 1968-1987
Road & Track on Jaguar 1950-1960
Road & Track on Jaguar 1961-1968
Road & Track on Jaguar 1968-1974
Road & Track on Jaguar 1974-1982
Road & Track on Jaguar 1983-1989
Road & Track on Lamborghini 1964-1985
Road & Track on Lotus 1972-1981
Road & Track on Maserati 1952-1974
Road & Track on Maserati 1975-1983
Road & Track on Mazda RX7 1978-1986
Road & Track on Mercedes 1952-1962
Road & Track on Mercedes 1963-1970
Road & Track on Mercedes 1971-1979
Road & Track on Mercedes 1980-1987
Road & Track on MG Sports Cars 1949-1961
Road & Track on MG Sprots Cars 1962-1980
Road & Track on Mustang 1964-1977
Road & Track on Peugeot 1955-1986
Road & Track on Pontiac 1960-1983
Road & Track on Porsche 1961-1967
Road & Track on Porsche 1968-1971
Road & Track on Porsche 1972-1975
Road & Track on Porsche 1975-1978
Road & Track on Porsche 1979-1982
Road & Track on Porsche 1982-1985
Road & Track on Porsche 1985-1988
Road & Track on Rolls Royce & B'ley 1950-1965
Road & Track on Rolls Royce & B'ley 1966-1984
Road & Track on Saab 1955-1985
Road & Track on Toyota Sports & GT Cars 1966-1984
Road & Track on Triumph Sports Cars 1953-1967
Road & Track on Triumph Sports Cars 1967-1974
Road & Track on Triumph Sports Cars 1974-1982
Road & Track on Volkswagen 1951-1968
Road & Track on Volkswagen 1968-1978
Road & Track on Volkswagen 1978-1985
Road & Track on Volvo 1957-1974
Road & Track on Volvo 1975-1985
Road & Track - Henry Manney at Large and Abroad

BROOKLAND CAR AND DRIVER SERIES
Car and Driver on BMW 1955-1977
Car and Driver on BMW 1977-1985
Car and Driver on Cobra, Shelby & Ford GT 40 1963-198
Car and Driver on Corvette 1956-1967
Car and Driver on Corvette 1968-1977
Car and Driver on Corvette 1978-1982
Car and Driver on Corvette 1983-1988
Car and Driver on Datsun Z 1600 & 2000 1966-1984
Car and Driver on Ferrari 1955-1962
Car and Driver on Ferrari 1963-1975
Car and Driver on Ferrari 1976-1983
Car and Driver on Mopar 1956-1967
Car and Driver on Mopar 1968-1975
Car and Driver on Mustang 1964-1972
Car and Driver on Pontiac 1961-1975
Car and Driver on Porsche 1955-1962
Car and Driver on Porsche 1963-1970
Car and Driver on Porsche 1970-1976
Car and Driver on Porsche 1977-1981
Car and Driver on Porsche 1982-1986
Car and Driver on Saab 1956-1985
Car and Driver on Volvo 1955-1986

BROOKLANDS PRACTICAL CLASSICS SERIES
PC on Austin A40 Restoration
PC on Land Rover Restoration
PC on Metalworking in Restoration
PC on Midget/Sprite Restoration
PC on Mini Cooper Restoration
PC on MGB Restoration
PC on Morris Minor Restoration
PC on Sunbeam Rapier Restoration
PC on Triumph Herald/Vitesse
PC on Triumph Spitfire Restoration
PC on VW Beetle Restoration
PC on 1930s Car Restoration

BROOKLANDS MOTOR & THOROGHBRED & CLASSIC CAR SERIES
Motor & T & CC on Ferrari 1966-1976
Motor & T & CC on Ferrari 1976-1984
Motor & T & CC on Lotus 1979-1983

BROOKLANDS MILITARY VEHICLES SERIES
Allied Mil. Vehicles No.1 1942-1945
Allied Mil. Vehicles No.2 1941-1946
Dodge Mil. Vehicles Col. 1 1940-1945
Military Jeeps 1941-1945
Off Road Jeeps 1944-1971
Hail to the Jeep
US Military Vehicles 1941-1945
US Army Military Vehicles WW2-TM9-2800

BROOKLAND HOT ROD RESTORATION SERIES
Auto Restoration Tips & Techniques
Basic Bodywork Tips & Techniques
Basic Painting Tips & Techniques
Camaro Restoration Tips & Techniques
Custom Painting Tips & Techniques
How to Build a Street Rod
Mustang Restoration Tips & Techniques
Performance Tuning - Chevrolets of the '60s
Performance Tuning - Ford of the '60s
Performance Tuning - Mopars of the '60s
Performance Tuning - Pontiacs of the '60s

BROOKLANDS BOOKS

CONTENTS

5	Lincoln Cosmopolitan	Car Classics	April	1973
20	The White House Carriages	Motor	April 4	1951
22	Lincoln Motor Trial	Motor Trend	July	1951
24	'52 Lincoln Road Test	Motor Trend	July	1952
28	Extensive US Ford Redesign	Autocar	March 14	1952
31	The 1952 Lincoln Road Test	Speed Age	June	1952
37	Lincolns with 205 bhp	Autocar	Dec. 26	1952
38	'53 Lincoln Preview	Speed Age	Jan.	1953
42	The 1953 Lincoln Road Report	Auto Sport	March	1953
44	America's Fastest Stock Car! Road Test	Road & Track	March	1953
46	Capri — '53 Lincoln Combines Luxury with Power Road Test	Motor Trend	March	1953
50	Lincoln 205 Capri Research Report	Speed Age	May	1953
54	A Refined Lincoln for 1954	Automobile Topics	Jan.	1954
55	Mexican Pan-American Race	Lincoln Div.		1953
64	'54 Lincoln Road Test	Motor Trend	Aug.	1954
66	With the Lincolns in Mexico	Motor Trend	Jan.	1955
71	Presenting the 1955 Lincoln Capri	Car Life	Dec.	1954
72	'55 Lincoln Capri Road Test	Motor Trend	May	1955
75	Thunderbird, Lincoln and Mercury Road Test	Auto Age	May	1955
82	'55 Lincoln Capri Road Test	Motor Life	June	1955
84	Lincoln Road Test	Science & Mechanics		1955
86	The '56 Lincolns	Motor Trend	Oct.	1955
92	Rebirth of the Continental	Car Life	Nov.	1955
99	The Safest Car in America	Car Life	Oct.	1955
102	Styling of the New Continental	Motor Life	Dec.	1955
107	Lincoln Capri	Car Life	Oct.	1955
108	Engineering of the New Continental	Motor Life	Dec.	1955
112	'56 Lincoln Road Test	Motor Trend	Dec.	1955
116	Lincoln and Mercury Comparison Test	Auto Age	Feb.	1956
122	New Continental Mark II Drivers Report	Motor Life	July	1956
126	'57 Lincoln — Top Driving US Car?	Motor Trend	Nov.	1956
132	The New Lincoln Road Test	Motor Life	Nov.	1956
134	The 1957 Lincoln Drivers Report	Motor Life	Nov.	1956
136	Lincoln — 1957	Car Life	Dec.	1956
137	First New Lincoln Continental Convertible in Nine Years	Wheels	Feb.	1957
138	Lincoln	Motor Trend	Jan.	1957
140	Lincoln Road Test	Motor Life	Feb.	1957
142	Lincoln Consumer Analysis	Car Life	May	1957
146	Lincoln 1958	Motor Life	Dec.	1957
148	Lincoln Consumer Analysis	Car Life	Jan.	1958
152	Lincoln Continental Mark III Road Test	Motor Life	March	1958
158	Continental Mark III Road Test	Road & Track	Aug.	1958
161	Lincoln for '59	Motor Trend	Dec.	1958
162	Lincoln, Imperial and Cadillac Comparison Test	Motor Trend	Sept.	1958
166	Lincoln	Motor Life	Dec.	1958
168	Lincoln Consumer Analysis	Car Life	May	1959
172	Lincoln — 1960	Motor Life	Dec.	1959
174	Lincoln for '60 Consumer Analysis	Car Life	April	1960
176	Testing the Luxury Cars Comparison Test	Motor Life	Aug.	1960

BROOKLANDS BOOKS

ACKNOWLEDGEMENTS

Brooklands are a small company who publish books, mainly on automobiles, for motoring enthusiasts. What is unusual about them is that they contain little or no new material.

The majority of our titles fall into our 'road test' reference series the main object of which is to bring together stories that when read in sequence present a comprehensive international picture of a model or as in this case a series of models produced by one manufacturer. Amongst our 350 titles is a second on Lincoln – Lincoln Continental 1961-1969 – which carries the Lincoln story on to the next generation of Continentals.

The worlds leading automotive publishers have supported the Brooklands series for over 30 years by allowing us to reissue their copyright road tests and reports. We are sure that Lincoln devotees will wish to join with us in thanking the management of Auto Age, Autocar, Automobile Topics, Auto Sport, Car Classics, Car Life, Lincoln Division, Motor, Motor Life, Motor Trend, Road & Track, Science and Mechanics, Speed Age and Wheels, for their understanding and generosity. Our thanks also go to David Lewis for letting us include his comprehensive article – The Gleam in Edsel Ford's Eye – which opens our story.

R.M. Clarke

Lincoln Cosmopolitan:

THE GLEAM IN
EDSEL FORD'S EYE

TAKING SHAPE — Part way through the styling process, the Lincoln fastback undergoes final touches on the clay body form. Note the lower grill looks much like the 1946-48 Ford front end.

STYLING FINISHED — With glass windows (or Plexiglass) in place, and tinfoil fitted over chrome parts, the completed clay mock-up is ready for approval by executives. (It was never accepted in this form.) Note the station wagon body in the background.

SCOOP GRILL — The Lincoln stylists studied the separated grill concept of front end design, but eventually rejected it in favor of the massive "water buffalo" mouth seen on the 1949-50 models. This clay model was a model of 1948, completed in 1943. Packard can be seen on the table.

THE GLEAM IN EDSEL FORD'S EYE

by David L. Lewis

If any automobile has reason to have an inferiority complex, that car is the '49 Lincoln Cosmopolitan. Lincoln chroniclers have endlessly sung the praises of the classic Lincolns and their coachbuilders of the '20's and '30's and of the later Continentals. But no Lincoln biographer has devoted more than a few passing paragraphs to the Cosmopolitan. Yet this stepchild of the Lincoln family, the first postwar-designed car to be introduced by Ford, was an exceptional automobile in its own right.

The Cosmo began, as did the Continental, as a gleam in Edsel Ford's eye. Although immersed in war work and battling the illness that would lead to his death in May, 1943, Edsel in late 1942 instructed designer Eugene T. Gregorie to sketch lines of Fords, Mercurys, and Lincolns for the postwar market. Strickly speaking, Edsel had no business telling Gregorie to work on civilian product designs, for the automobile industry was supposed to be devoting all of its attention to the war effort. Not until September, 1944 did the War Production Board authorize auto manufacturers to begin work on postwar models. Nonetheless the enthusiastic Gregorie was soon grinding out sketches which, in turn, were being translated into clay models by early 1943.

At the same time Edsel sent Gregorie and his designers to their drawing boards, he also asked engineering executive Laurence Sheldrick to conduct research and make recommendations on chassis and mechanical improvements which could be incorporated into postwar cars. Sheldrick focused on independent suspension systems. His research was supplemented by that of a separate team of engineers at the Lincoln plant (located several miles east of the Engineering Building), which concentrated on torsion bar suspension.

In directing Gregorie and Sheldrick to style and study postwar configurations, Edsel suggested only that the postwar cars have clean, simple designs. Designers and engineers were, therefore, able to start with a fresh sheet of paper. In addition, Fordmen at this stage had only to please themselves and Edsel. Unlike General Motors, Chrysler, and most other automakers, Ford had no elaborate committee system to pass judgment on designs and layouts. Of course company personnel in the end might have to contend with a personality more difficult than any half-dozen committees — Henry Ford, now 80 years of age and becoming increasingly stubborn, and even senile.

Until Edsel's death, numerous designs, layouts, and clay models for postwar cars poured out of Gregorie's studio and Sheldrick's shop. But with the administrative vacuum created by Edsel's departure, postwar planning ground almost to a standstill.

The existence of the postwar designs and models was known only to a handful of company executives and technicians. Foremost among the former were manufacturing bosses Charles E. Sorensen and Peter E. Martin and purchasing executive A. M. Wibel. While Edsel was alive Henry Ford was only vaguely aware of the developmental work, never having attended meetings at which the postwar models were discussed nor having visited the quarters in which design and mechanical work was going on. But he soon would take a very personal, albeit short-lived, interest in the project.

Old Henry's interest in the postwar models was prompted by his grandson's interest in them. Twenty-five-year-old Henry II had been released from the Navy in August, 1943 to help his grandfather run the company. Nosing about, he learned in early September of his father's plans to design and engineer cars for the postwar market. Asking Gregorie and Sheldrick about the plans, he was told that Edsel had "talked quite a little to us about . . . the matter of postwar cars," then was shown designs and layouts. Young Henry, already more interested in the postwar civilian market than in war work, was pleased with what he saw.

Henry Ford, informed of his grandson's contacts with Gregorie and Sheldrick, was furious that Young Henry had been exposed to ideas not Old Henry's own. The Old Man ordered Sorensen to chew out Gregorie, and the manufacturing boss did so with a vengeance. Sheldrick fared worse; he was fired after a heated exchange with Sorensen. Years later Sheldrick theorized that "if there was any postwar designs to be created, Henry Ford wanted to be the dictator of what those designs were to be, and he knew that anything cooked up between Edsel and Gregorie and between Edsel and myself would have been somewhat of a progressive, modern nature. He wasn't in the mood for accepting any of the modern innovations such as independent suspension and some of the so-called approaches to stream-lining which was the vogue and talk of the day."

At this juncture the steam would have gone out of postwar product planning had it not been for Henry II. Made a vice president in December, 1943 and executive vice president in April, 1944, Young Henry — as his grandfather faded from the scene and Old Henry's hatchet man, Harry Bennett, felt it politic to stay out of Henry II's way — insisted that advanced product planning be continued. In early June, 1944 Henry II, sales manager John R. Davis, and production executives Ray R. Rausch and Mead L. Bricker (Sorensen himself was fired by Old Henry in

OLD LINCOLN — In 1942, when Edsel Ford authorized the start of a new Lincoln body style for the postwar market, this clay model was begun. The 1948 Lincoln shown here, created no great stir in the market place, but Edsel Ford hoped his new car would capture the American public's attention.

FAST BACK — Ford stylists und[er] Eugene T. Gregorie worked on a fa[st] back body for the new Lincoln, a[...]

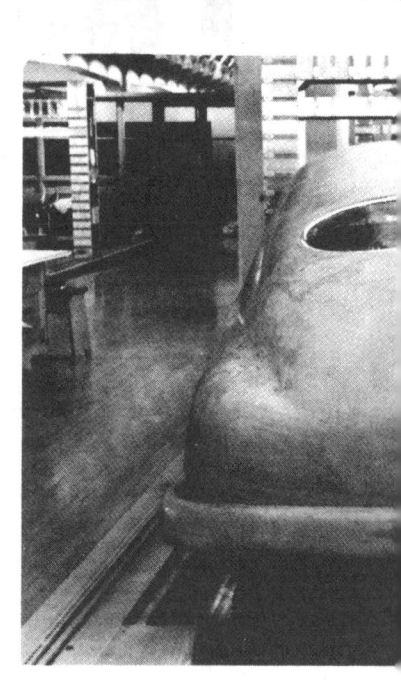

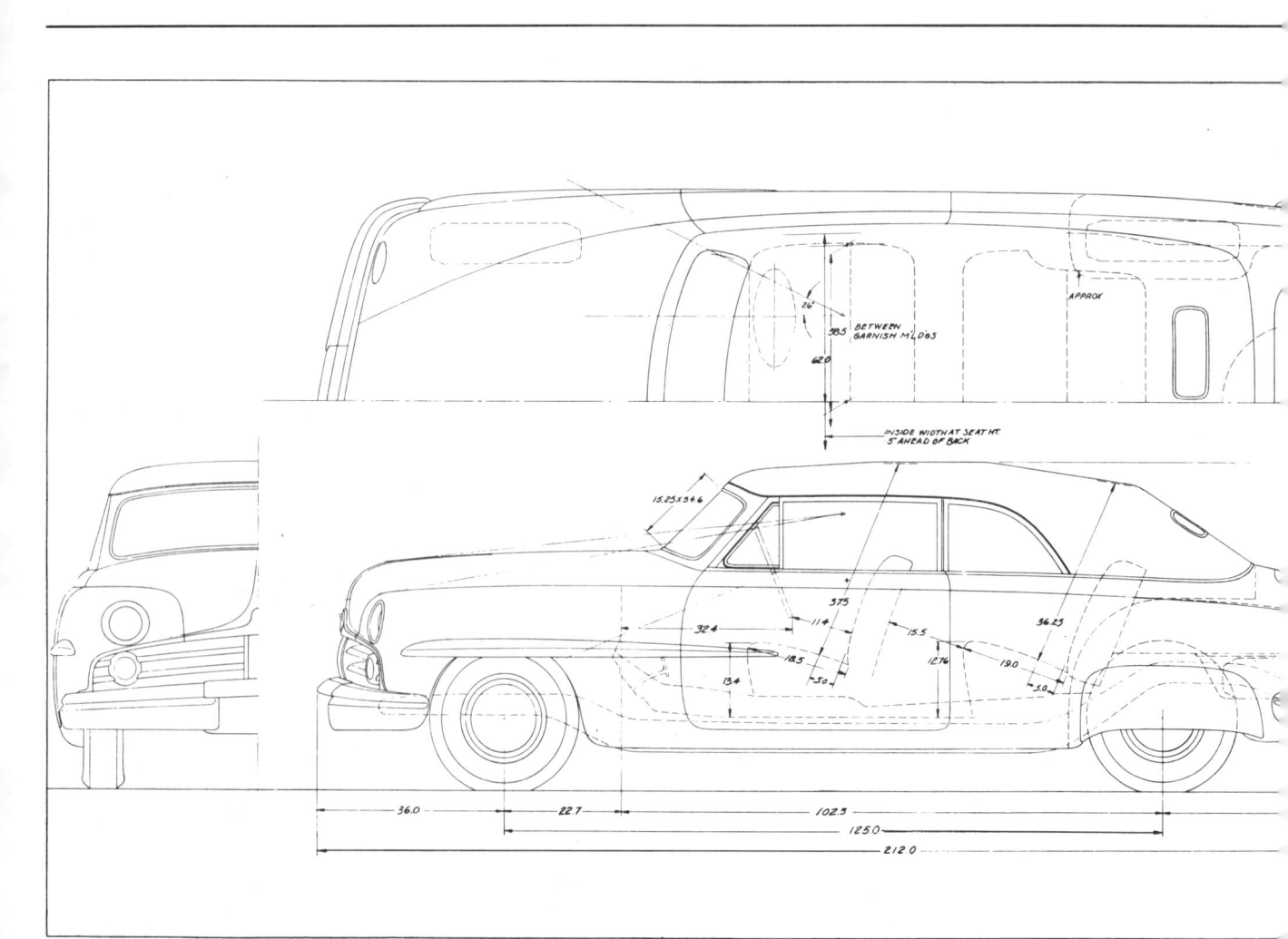

his photo shows the direction of their thinking. Note only one tail light is finished.

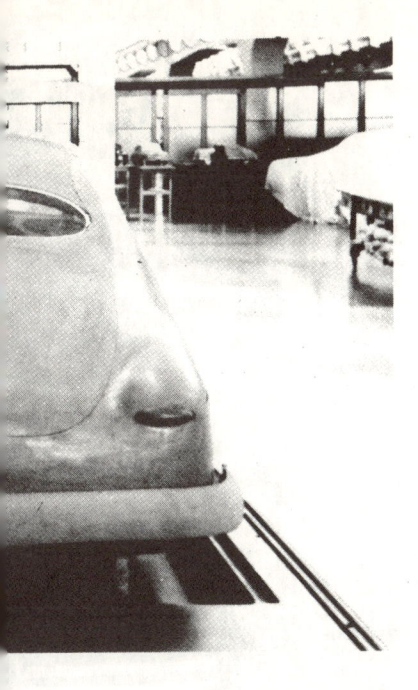

FUTURISTIC FRONT — Experiments with the new body led to front end designs such as this one which features a two-part grille. Oldsmobile in 1956 picked up on this same theme with its "scoop" grille of that year.

ROUNDED FRONT — While lacking in excitement, this early design experiment proved that Gregorie had a free hand in designing a completely new Lincoln. This general body style was retained for the final fastback put into production, but the front was changed.

BULKY LOOK — The first 1943 front end treatment resulted in a bulky looking automobile and was probably dropped for that reason. (Above center)

MECHANICAL DRAWINGS — The proposed new body was carefully studied in hundreds of drawings like this one, which give a clear idea of the size and shape of the Cosmopolitan convertible.

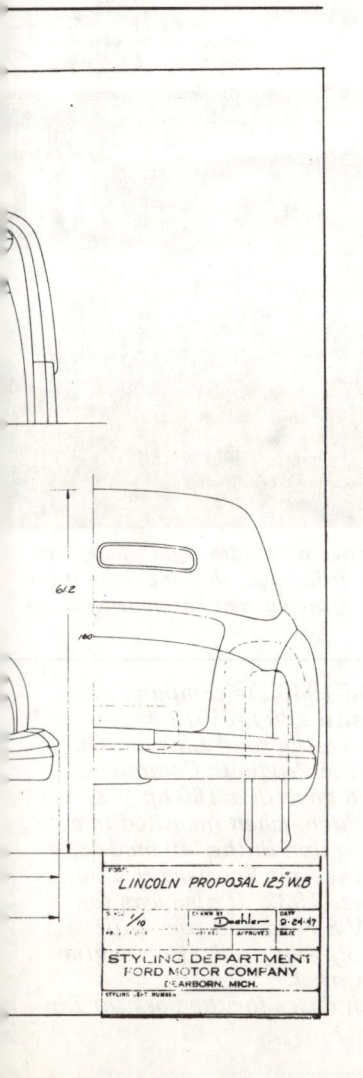

9

early 1944) formed an Engineering Planning Committee to chart postwar product strategy. Members of the committee included the four organizers plus Gregorie, Charles H. Carroll, director of Purchasing; Richard Kroll, head inspector; William F. Pioch, head of Tool Engineering; R. H. McCarroll, head of Engineering, Jack Wharam, an Engineering executive; and V. Y. Tallburg, an engineering coordinator who also served as secretary of the group.

Eighty-one-year-old Henry Ford was not a member of the committee, nor did he, as Tallburg observed later, take any part in engineering by 1944. "Once in a while," noted Tallburg, "if we wanted to make a major change in the engine or something like that, we would show it to him. But he didn't take much interest."

The Engineering Planning Committee and its successor, the Products Committee, formed under the chairmanship of John R. Davis in December, 1945, guided design and developmental work on Ford's car lines through the remainder of the war into the postwar era. The Lincoln program occupied less committee time than the Ford and Mercury programs, which in those days, to use Sheldrick's words, were "put in the same basket," while the Lincoln was put in another basket. The committee's main concern for several years was the company's projected light-car program, specifically whether to build such a car, and, if so, what size, and when.

In contrast to the backing and filling associated with the light-car program, plans for the postwar Lincoln proceeded with relative smoothness from start to finish. Everyone took it for granted that newly-designed Lincolns would bow in the postwar era; the key question, aside from matters of design and mechanics, was when.

The Lincoln program was abetted by broad philosophic agreement among those chiefly concerned — Edsel, Gregorie, Sheldrick, Henry II, John R. Davis, and later, Harold T. Youngren, an ex-General Motors and Borg-Warner engineering executive who became a Ford vice president and director of engineering (and styling) in August, 1946. All saw the need for smoother, flowing lines, lower overall height, and general clean-up of design. All envisioned wider bodies and more window area and pushing the engine forward to cradle passengers between the front and rear axles for a better ride. All also saw postwar automobiles "as more and more becoming a comfortable room-on-wheels instead of just a place where people sit while being carried from one point to another." This concept was to be translated into wider seats by merging fenders into body lines and eliminating running boards.

The engine program elicited less agreement among the principals.

ANOTHER TRY — Still using the clay form, Gregorie tried a "hidden light" look, with the grille treatment more conventional. Like the two-grille idea, this one looks a bit pudgy from the front. Note Cadillac in background. All of the current competition auto-

Until mid-1946 the company planned to use a 120 hp V-12 engine in the down-the-road Lincoln. On June 13 the Products Committee elected to go with a 160 hp V-8 motor, which, when installed in a 152 hp version in the '49 models, became the first V-8 used in Lincolns since 1932. It also was the largest V-8 Ford had ever built and the biggest production engine on the market.

Launch dates for the postwar Lin-

mobiles were checked against the proposed Lincoln design. There is a 1949 Ford station wagon in the background as well.

ducts Committee recommended that all postwar Ford products have independent front suspension (as opposed to spring suspension) and semi-elliptic rear springs. The Lincoln's wheelbase was set at 123 inches; Mercury at 120; Ford 117; and the proposed light car 98. Committee members wondered for some months whether the postwar Mercurys and Fords might be too large for the market. But they consistently saw merit in a large Lincoln, and, on May 10, 1945 increased the wheelbase of the '49 Lincoln line from 123 inches to 125 inches.

Body styles for the Lincoln evolved smoothly in Gregorie's studio. The program was greatly aided by the close supervision of Henry Ford II and John R. Davis, who prior to executive vice president Ernest R. Breech's arrival in mid-1946, had more to say about Ford's postwar products than any other executive. Both Young Henry and Davis were regular, at times almost daily, visitors to Gregorie's studio.

Thinking and progress on the '49 Cosmo and its smaller stablemate, simply called the Lincoln, is easily traced through the minutes of Ford's top-level committees. Chassis construction got underway in late 1945; also sheet metal mock-ups of the front end. Cost analysis was a recurring theme; also the performance of various components. Complaints were heard, for example, in April, 1946 that the hydraulic-operated windows were occasionally "freezing."

On July 10, 1946, at a meeting of the newly-formed Engineering-Management Committee, Breech suggested that fluid drive be considered for the postwar Lincolns. It was; but neither it nor any other automatic transmission found its way into the '49 products. At a September 3, 1946 Engineering-Management meeting, much of the '49 model program was changed. The smaller Lincoln emerged at this point, its wheelbase being dropped from 125 to 121 inches, while the future Cosmo's wheelbase retained the longer length. The Lincoln (but not the Cosmo) — and also the Mercury — was to use the so-called 7-A body (designed originally by Gregorie as the postwar Ford) and its front end, fenders, hood, and other metal parts were to be re-designed to match the 7-A body. The front end was to be adapted to make use of the same grille as that on the Cosmo.

coln and other Ford lines were changed repeatedly. Henry Ford II, inexperienced and enthusiastic, wondered aloud at a December 16, 1944 Engineering Products Committee meeting whether the vehicles could be introduced in 1945. Tactfully, veteran Fordmen replied that much more lead time would be required. At first committee members referred to the postwar lines as '47 models, then as '48's. The cars finally were introduced as '49's in April, 1948. Extra time was needed to develop the three postwar lines. Beyond this consideration, there was no burning competitive need to bring out newly-designed cars before 1948, given the seller's market of the late 1940's.

On December 16, 1944, Young Henry also decreed that the company's postwar lines would have "definite distinctions." Eleven days later, the Engineering Pro-

11

1949 LINCOLN ENGINE

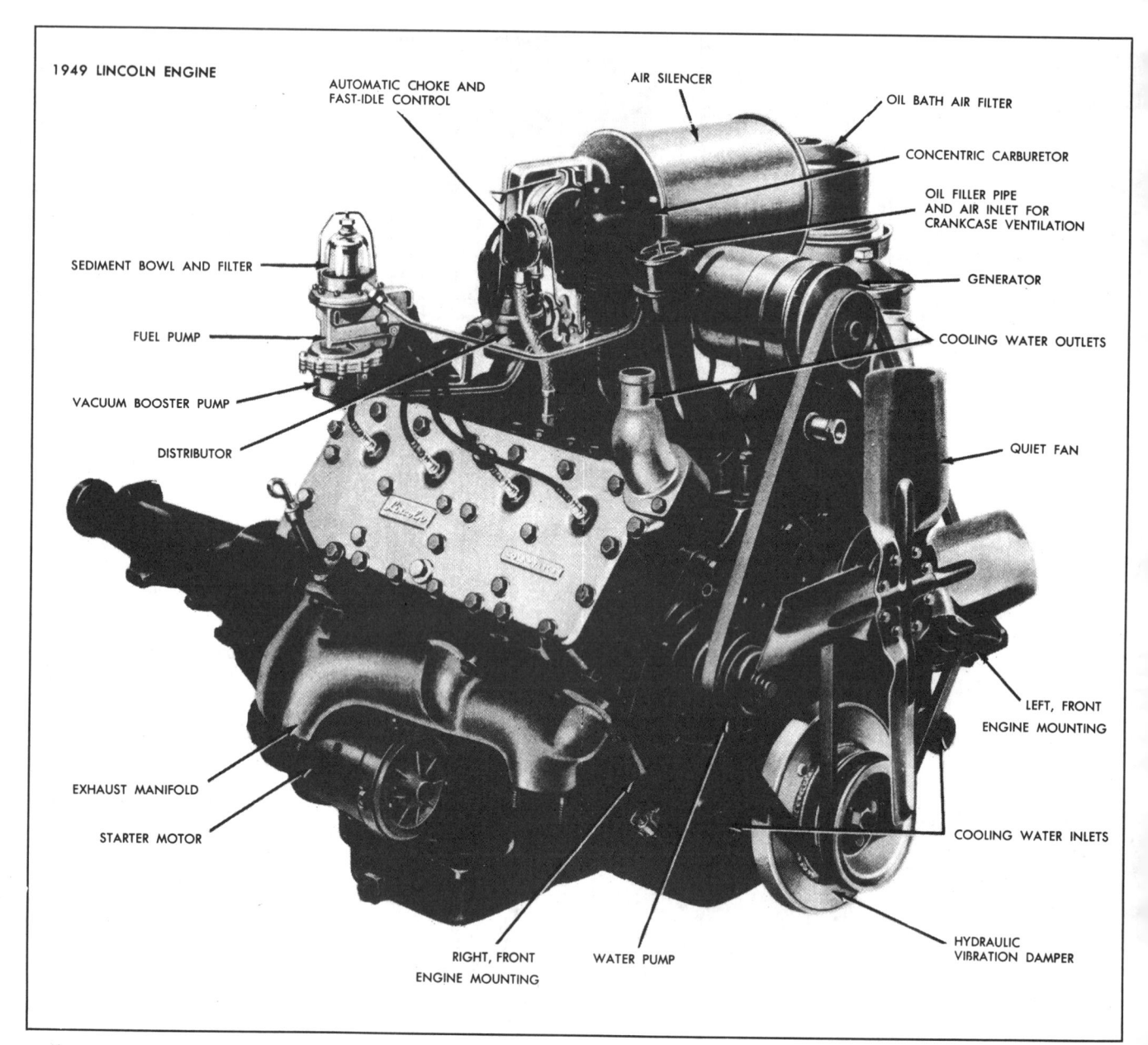

Labels: AUTOMATIC CHOKE AND FAST-IDLE CONTROL; AIR SILENCER; OIL BATH AIR FILTER; CONCENTRIC CARBURETOR; OIL FILLER PIPE AND AIR INLET FOR CRANKCASE VENTILATION; SEDIMENT BOWL AND FILTER; GENERATOR; FUEL PUMP; COOLING WATER OUTLETS; VACUUM BOOSTER PUMP; DISTRIBUTOR; QUIET FAN; LEFT, FRONT ENGINE MOUNTING; EXHAUST MANIFOLD; STARTER MOTOR; COOLING WATER INLETS; RIGHT, FRONT ENGINE MOUNTING; WATER PUMP; HYDRAULIC VIBRATION DAMPER

NEW ENGINE — Adapted from a truck motor, the 1949 Lincoln motor was the biggest production car engine built that year in the U.S. While it had lots of torque, it lacked the snap of Cadillac's smaller overhead valve V-8, a fact that was quickly apparent to power concious Lincoln owners.

Thus by the early fall of 1946 the postwar Lincoln program was well along. By that time executives had decided to use a new "K" type frame instead of the former unit body construction, a new eight-cylinder engine to replace the 12-cylinder plant, torsion bar front suspension instead of front cross spring, a new side spring Hotchkiss drive instead of torque tube, a new larger transmission and automatic overdrive. At this point, there also was hope that a new automatic transmission being developed by Detroit Gear would be available on the '49's. Testing on most of these components was well along by the fall of 1946, twelve '46 Lincolns having been equipped and checked out with torsion bar suspension and V-8 engines. Also available for testing as early as July, 1946 were two Cosmo prototypes. By fall three more Cosmo prototypes in four body styles were running on the Dearborn test track and nearby country roads.

Instrument panels and steering wheels were selected at the January 13, 1947 Products Committee meeting; trim fabrics at the March 10 gathering. In May, 1947 the Cosmo's body program received final approval: a four-door sport sedan, a four-door fast-back town sedan, a six-passenger coupe (which looked more like a two-door sedan), and a six-passenger convertible. The fast-back, as it turned out, was offered only during the '49 model year. Thus by mid-1947 the Cosmo's design was essentially set. It remained only for executives to make such decisions as engine color (blue) and whether to include such features as automatic seat adjustment as standard or optional equipment (they went with standard); virtually all of these decisions had been made by October, 1947.

To expedite Lincoln design and production, the Ford Company established a Lincoln Division in 1945 (combined into Lincoln-

Lincoln transmission and overdrive

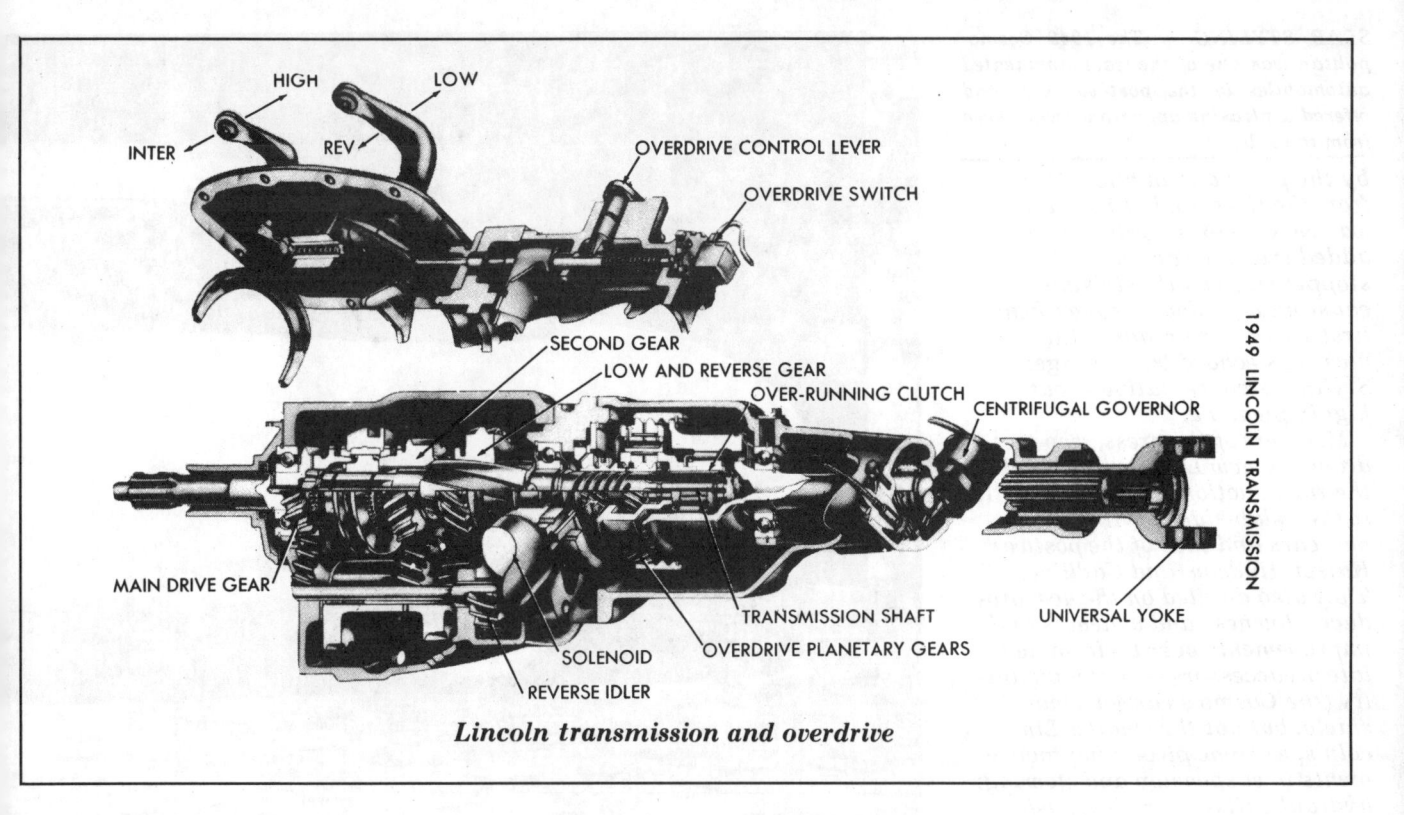

NICE OVERDRIVE — *The 1949 Lincoln unit was one of the best ever put in an American automobile. This was used in place of an automatic transmission in the Cosmopolitan series in 1949.*

BIG CLUTCH — *To withstand the massive torque of the new Lincoln V-8, a big clutch was created, which according to contemporary reports, was found entirely satisfactory by Lincoln owners.*

Mercury Division in 1946) and rapidly recruited a talented team of engineers, technicians, stylists, and artists. Upon the retirement of Lincoln's chief engineer, Frank Johnson, in 1947, Earle S. MacPherson was brought in. Later MacPherson was named the company's chief engineer in charge of research and engineering. The Division also modernized and installed new machinery in parts plants and built three new assembly plants — in Los Angeles, St. Louis, and Metuchen, New Jersey. All in all, the Division invested $75,000,000 in the '49 Lincoln and Cosmopolitan. At the same time it built a network of exclusive Lincoln-Mercury dealers, who it is estimated, also spent another $75,000,000-plus on buildings and equipment.

The Cosmo and smaller Lincoln were unveiled on April 22, 1948, a week in advance of the Mercury's debut, seven weeks ahead of the Ford's introduction. The Cosmo, now Lincoln's top-of-the-line product (the Continental having been "temporarily" discontinued on March 30), was favorably received

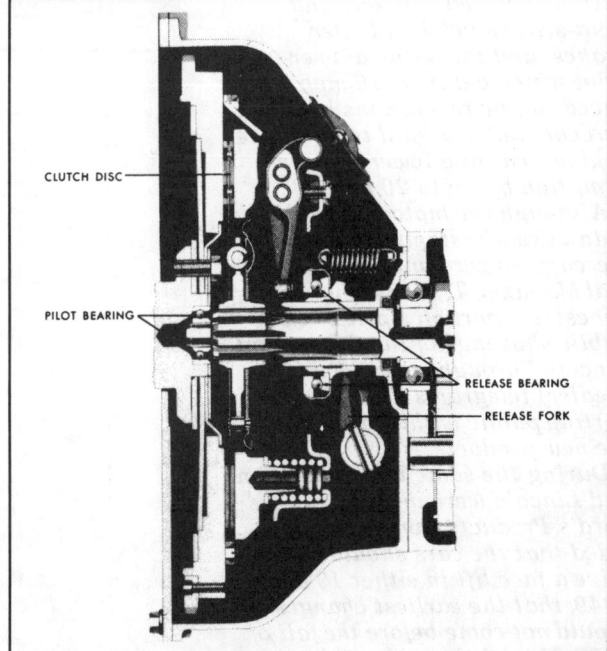

1949 LINCOLN TRANSMISSION   F1-F2

The Power Train

F1—The "power train" is made up of those parts of the chassis which deliver power from the engine to the rear drive wheels—the clutch, transmission, overdrive (if installed), propeller shaft assembly with front and rear universal joints, rear axle assembly with axle shafts, wheels and tires, and the parallel rear springs to which the axle assembly is mounted. The Lincoln "power train" is shown in the diagram.

F2—CLUTCH: The easy action of the new Lincoln clutch results from use of specially weighted clutch release levers. Centrifugal force produced on these levers by the rotation of the clutch supplements the pressure exerted by the clutch springs. This force increases with engine speed, thus decreasing possibility of clutch slippage at higher speeds.

SLAB STYLING — The 1949 Cosmopolitan was one of the least ornamented automobiles in the post-war era, and offered a pleasing appearance when seen from the side.

by the press and public. "New from the tires up, but not radical," was Newsweek's verdict, which added that the car was an "eye-stopper sufficiently striking to cause many a double-take when first seen on the roads." This view was echoed by Iron Age: Styling is conservative — but highly pleasing."

Members of the press, who test-drove the new Lincolns the week of the introduction, noted similarities in the "slab-side" styling of the new cars and that of the postwar Kaiser, Hudson, and Cadillac. They also dwelled on the new products' lowness and wideness and improvements over their immediate predecessors — better visibility (the Cosmo's curved windshield, but not the smaller Lincoln's, was one-piece); improvements in suspension and steering; hydraulically-controlled, push-button operated windows and front-seat movement; better brakes; and the optional overdrive which cut in at 26 mph, reduced engine revolutions by 23 percent, and was said to minimize engine wear and lower gas consumption by up to 20 percent.

Although the motoring public didn't exactly stampede Lincoln-Mercury showrooms, Division General Manager T. W. Skinner could honestly report on April 23 that public showings of the Cosmo and Lincoln "brought in a flood of (dealer) telegrams Thursday reporting public enthusiasm" for the new products.

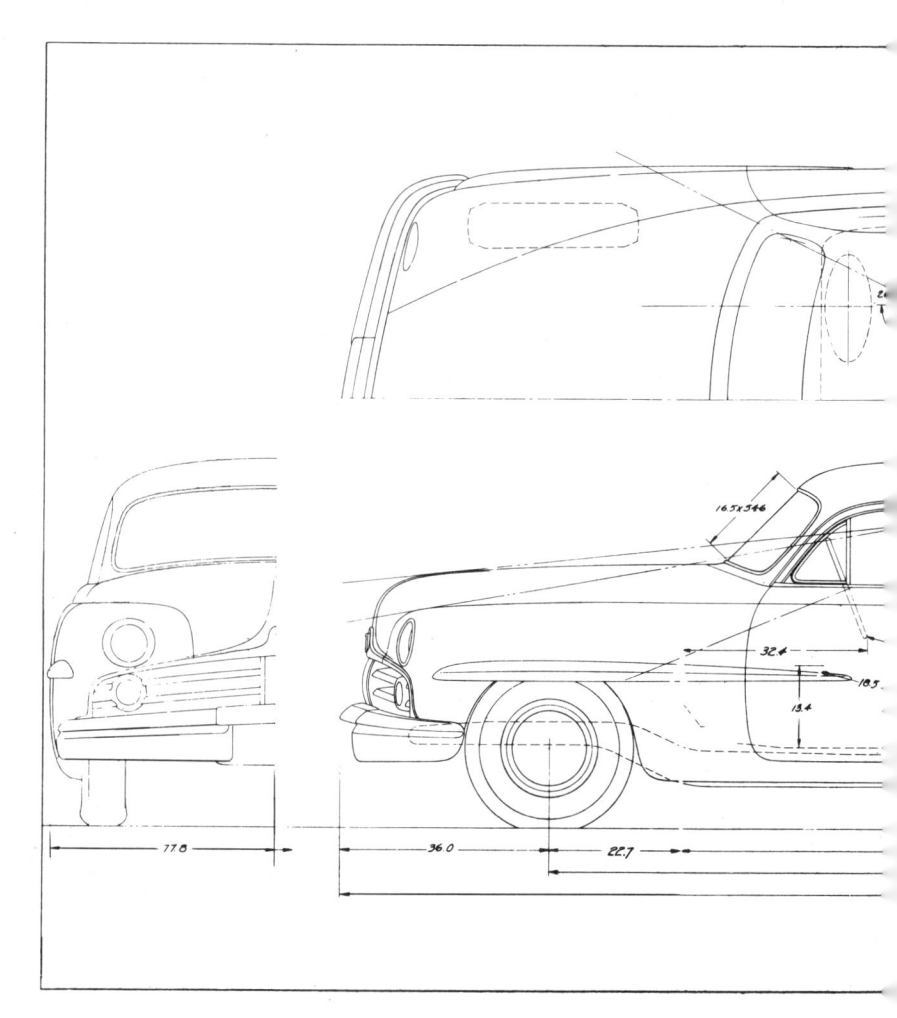

During the same week the Cosmo and Lincoln were introduced, Ford's Products Committee decided that the cars should not receive a face lift in either 1948 or 1949; that the earliest changes should not come before the fall of 1950. Nonetheless, throughout 1948 Ford executives wrestled with decisions on how to deal with complaints and make improvements on the vehicles. They hoped eventually to incorporate a one-piece rear window on the Cosmo, provided it could be produced at less cost. They also wanted an improved instrument panel; the old one was considered "drab in appearance," particularly insofar as the ash tray and glove compartment door were concerned. Some also thought that the row of switches beneath the radio looked too much like an organ keyboard.

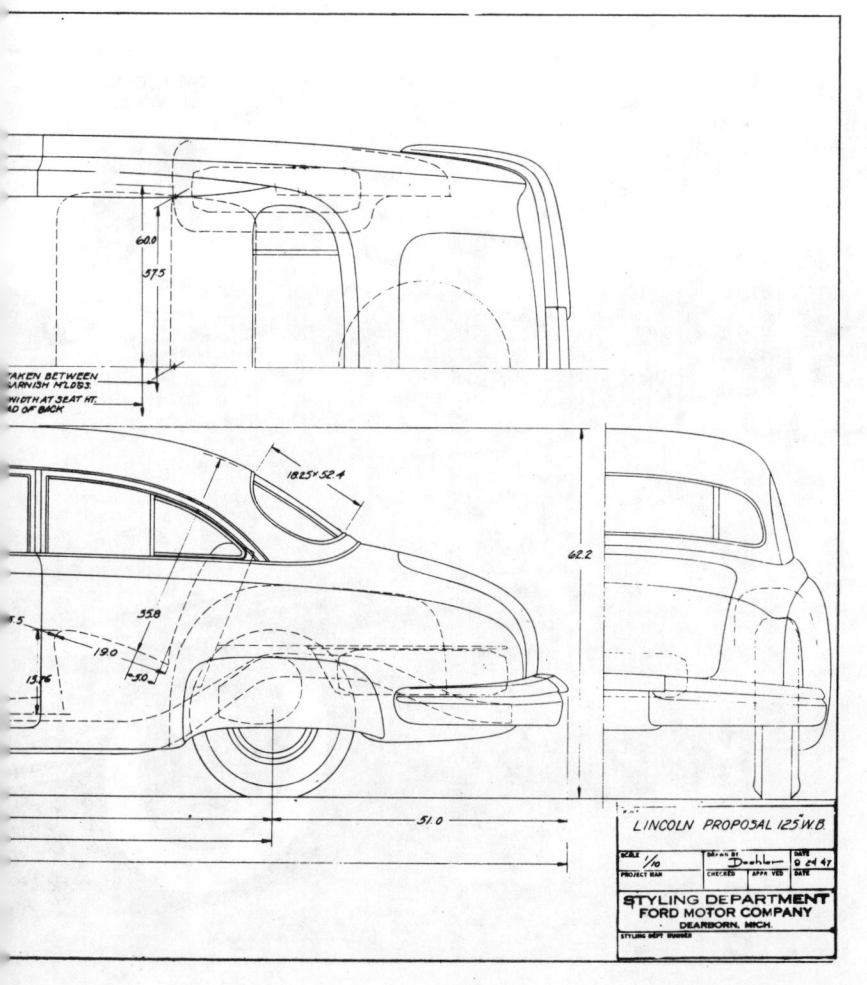

In July, the Products Committee considered changing the hinging on the Cosmo's doors to conform to hinging throughout the industry. Such a change was regarded as too expensive to make before a complete body overhaul three years into the future; but a tooling estimate was authorized. At the same meeting the committee considered a variety of defects that had surfaced in the new cars. The fuel pump on the Cosmo was too close and scraped the firewall; the Cosmo's floorboard screws were too long and hit the engine, making an annoying noise; the hydraulic cylinders in the Cosmo's doors were loose, causing more noise; dust entered the Cosmo's trunk compartment between the drip moulding and the chrome strip; the handbrake didn't hold even on slight inclines; hardware rattled; there was insufficient space between tire and wheel housing for mounting tire chains, etc.

On December 30, 1948 the Products Committee again thoroughly aired Lincoln defects, and members agreed on a long list of down-the-road styling changes. Everyone agreed that the grille was unsatisfactory in appearance; and Engineering was instructed to prepare styling sketches showing a grille with the cross bars running straight instead of slanting or "drooping" as on the '49's. New better-looking bumper guards were regarded as a must, as well as a new hood ornament. The group also insisted that continuing door rattles be eliminated, and told the engineers they should also come up with a door that slammed with the "quality sound desirable in this price class automobile." The steering wheel was said to be stiff; also, more insulation between the steering column and instrument panel was required. Later it was found that a a weak shift lever hinge on the steering column occasionally caused the shift lever to snap off in the driver's hand. Vacillating between a desire to make improvements and a reluctance to spend money on them, the company actually made only grille and trim changes and tinkered with annoying minor problems that wouldn't go away unless resolved.

DETAILED PLANS — The dimensions of the Cosmopolitan were shown in detail in this drawing, which gives the cars height as 62.2 inches. (Bottom center.)

16

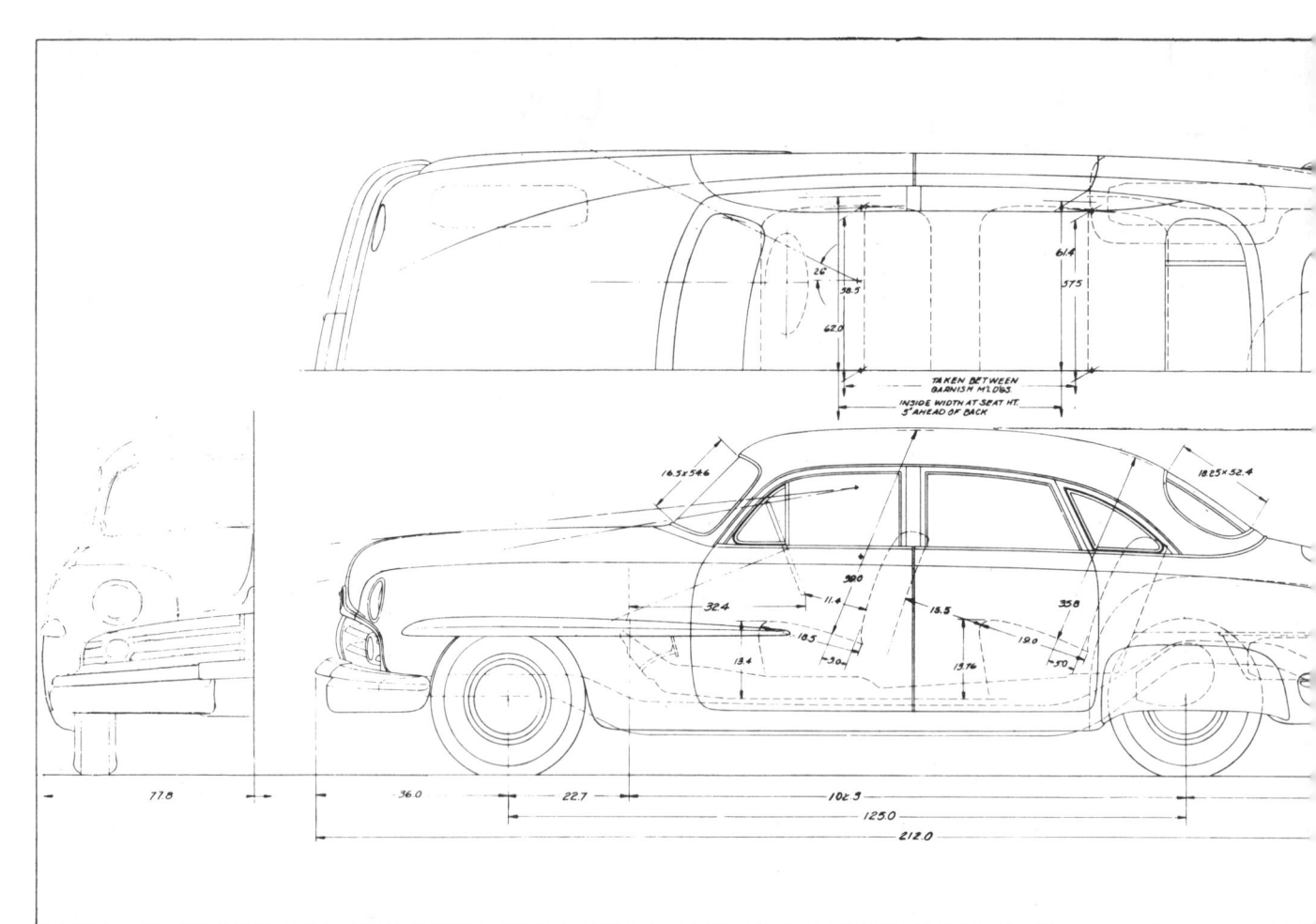

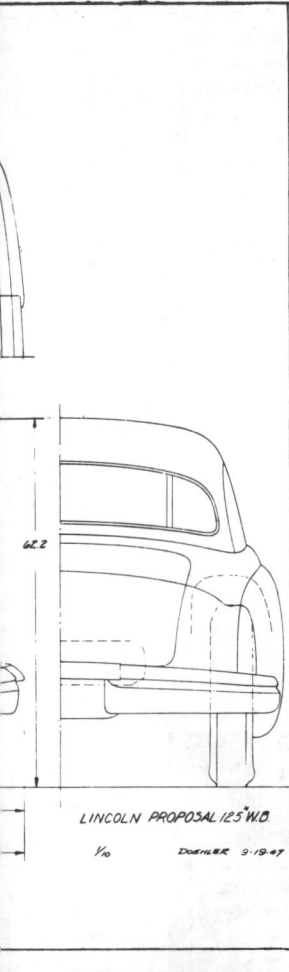

*FINAL DRAWINGS —
Finished by Bob Doehler
in 1947, this mechanical
drawing closely resembles
the 1949 Lincoln four-door
sedan. Doehler also made
similar detail sketches of
all the American luxury
cars of the period, Packard, Cadillac, Chrysler,
DeSoto and Studebaker so
that the 1949 Lincoln concepts could be compared.*

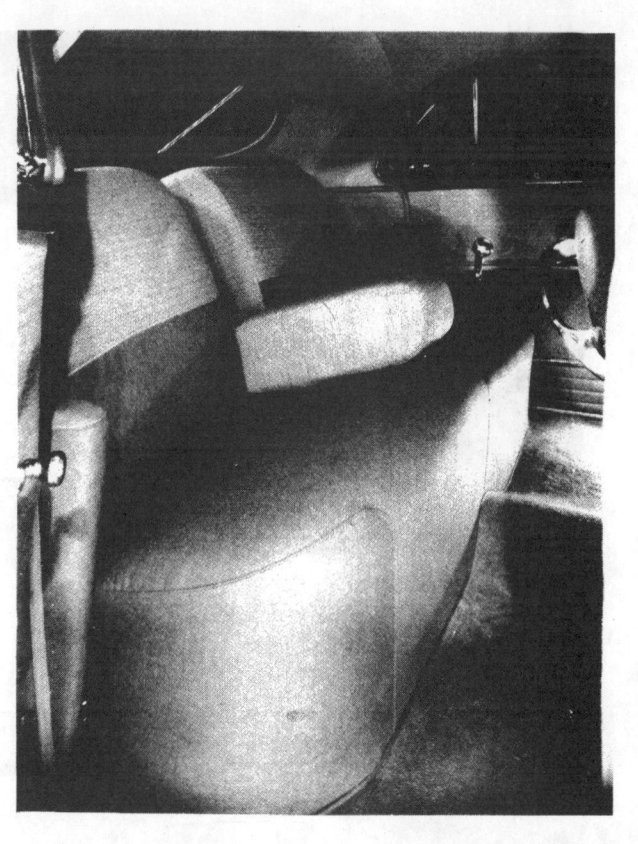

POSH INTERIOR — The interior of the Cosmopolitans for 1949 were finished in a material called Grey Birdseye broadcloth. Note the deep pile carpets and fold down arm rest for extra comfort.

FASTBACK COSMO — The fastback idea finally saw sheet metal with this four-door sedan, but was probably not as popular as Edsel and Gregorie had originally hoped.

The most popular car in the Cosmo series proved to be the four-door sport sedan, selling for $3,238 and weighing in at 4,527 pounds. Most expensive of the four models was the convertible, priced at $3,948 and weighing 4,717 pounds. The club coupe cost $3,185 and weighed 4,487 pounds; the town sedan, $3,328 and 4,631 pounds.

It was unfortunate that a car designed from scratch should remain in production only three years. Ford retained the name (as a running mate to the Capri) in its '51 models. But the Cosmo, as well as the Capri, was offered on a 123 inch wheelbase, and both models used an all-new overhead valve V-8 engine and a new ball-joint front suspension developed by Lincoln and Thompson Products.

The '49-'51 Lincolns sold well. A total of 32,638 lines were registered in 1948, an all-time high. Sales increased to 37,691 in 1949, then fell back to 34,318 in 1950 as the Korean conflict got underway. Until 1965, the 1949 sales mark was topped only in 1953 and 1956.

The Cosmo gained some contemporary fame as well as moderate fortune. Ten oversized Cosmo Limousines were added to the White House fleet in 1950. One of them was returned to Ford, at President Eisenhower's request in 1954 to be revamped with a plexiglass roof. Afterwards it was known as the "Bubbletop." But today, as noted earlier, the Cosmo has been virtually relegated to the dustbin of Lincolniana — a seldom-mentioned skeleton rattling in the closet. Histories of the Lincoln say only a few words about it, one shows nary a picture of it, others only a few prints. The 1949-51 Cosmopolitans deserve better.

GETTING CLOSE — This clay mockup body is a close representation of the final form the 1951 Cosmopolitan's took. The sides, hood and right tail light are almost identical to the final car, while the grill treatment and bumper were vetoed. This is a 1945 clay rendering.

SMALL MERCURY — This mockup looks almost identical to the 1949 Mercury — a big contrast to the body style parked next to it which was dropped in 1948. This model was made in 1944.

ADVANCED DESIGN — Finished sometime in 1946, this advanced design for the new Lincoln was not used, but some minor details appeared on other Ford products, namely the Mercury which used part of the lower grille in modified form.

LINCOLN

LINCOLN-MERCURY DIVISION
of Ford Motor Company
Detroit, Michigan

6-Passenger Coupe

Cosmopolitan 6-Passenger Coupe

Used for fair-weather State functions, the convertible model of the fleet is beautifully finished. Flashing red lights and a siren are concealed in the grille.

IT has always been the practice for various American fine car manufacturers to keep the White House well supplied with long, luxurious limousines. Until recently the Presidential stable has been a mixed one. Cadillacs rubbed fenders with Packards, Pierce-Arrows, and Lincolns in the White House garage.

However, Lincoln-Mercury Division of the Ford Motor

The White House Carriages

Details of a Fleet of Ten Lincolns with Coachwork Henney which have Recently been Delivered to

Company corrected this un-American situation by furnishing President Truman with a complete fleet of special 145-inch wheelbase Lincolns. These cars are leased to the government for a nominal yearly sum, the title remaining with the company.

Nine of these cars are closed limousines, and the flagship of this group is shown in some of the accompanying photographs. The Henney Motor Company, whose normal speciality is the construction of hearses and ambulances, supplied the coachwork and it can be seen that this firm is as well aware of the requirements of distinguished and healthy living persons as it is of the sick and the dead.

Comprehensive Fittings

All metal fixtures in the passenger compartment of the No. 1 White House limousine are gold-plated. Special, fitted cases of brown lizard skin recessed into the arm rests hold Thermos bottles for coffee and water, a writing portfolio with gold pen and pencil, and a cigar and cigarette humidor. Auxiliary seats fold up against the partition behind the driver. Rear windows and the glass partition separating the two compartments can be controlled by push buttons located above the left arm rest. A radio control panel for the rear compartment is located above the right arm rest adjacent to the intercommunication system microphone. Upholstery is grey, shadow-stripe broadcloth with grey-grain, metal garnish mouldings offset by the gold-plated fixtures.

The driver's compartment is somewhat less luxurious, but equally well thought out. It includes a separate radio

Coachwork on the open car is by Dietrich, well known for fine body building in the U.S.A. The interior is upholstered in cherry-red and black leather. The photograph above shows the layout of this very roomy convertible.

Good proportions have been cleverly retained on a car of 20 ft. overall length mounted on a 145-in. wheelbase. At the rear can be seen the step and hand-hold for the secret-service bodyguard.

Flagship of the fleet is limousine No. 1, which has all interior metal parts gold-plated. Visible here is the large running board on which secret-service men stand when accompanying the President.

and heater, the standard Lincoln instruments and a slot under the seat for an umbrella. The latter item is undoubtedly useful as, nowadays, in Democratic Washington one cannot always count on an umbrella-equipped concierge being present at official destinations.

The tenth car, a convertible sedan, has coachwork by Dietrich, an old and honoured name in the custom body field. Details of trim are reminiscent of earlier days when American coachwork could and did compete favourable with anything the world had to offer. The upholstery is cherry-red and black leather and, considering its size, the cloth top folds back rather neatly. As in the closed cars, two fresh-air heaters, one under the bonnet and one fitted into the trunk, supply ventilation and heat to both front

Dietrich and   By Donald

resident Truman   MacDonald

and rear seat passengers. The rear heater is connected to the radiator with tubes, and a special fresh air inlet is visible on the top of the rear deck.

All of the cars are painted black and have hand-holds and extra-large running boards for secret-service men, who, in America, cling to the Presidential car wherever it goes. Warning sirens and flashing red lights further convince the onlooker that someone important is passing by.

The mechanical details of the cars are similar to the production Lincoln Cosmopolitan. The engine is the standard 152-horsepower V-type eight-cylinder unit which powers all Lincoln cars. However, a special heavy-duty Hydra-Matic transmission is fitted due to the great weight

(*Right*) The driver's compartment, separated by an electrically controlled glass division, has an individual radio and heater.

(*Below*) Space in the boot for Presidential luggage is sacrificed in favour of spare wheel, heater and intercommunication equipment.

(*Below*) Equipped with radio, intercommunication microphone, recessed refreshment cases and a writing-portfolio, the luxurious rear compartment of the No. 1 limousine seats five in comfort.

(6,450 pounds in the case of the convertible), and unusually large tyres are mounted. The cars are bulky, even by American standards. The convertible has an overall length of 20 feet and a width of over six and a half feet.

Despite their great size, the cars are exceptionally graceful. Lincoln designers and engineers, working under the direction of Harold Youngren (Vice President-Engineering) are to be commended on their solution of a difficult design problem, for the cars combine modernity with size and dignity.

These qualities are certainly desirable for Presidential transports, although White House officials and the American public were recently disconcerted to learn that the Sheik of obscure and far away Kuwait had taken delivery of what is undoubtedly the biggest official motor car in the world—a 163-inch wheelbase Cadillac.

LINCOLN MOTOR TRIAL

LARGE CARB housing (above), provides for air circulation around float chamber, prevents over-heating of fuel. Viland verified small jet size

INSTRUMENTS ARE positioned faultlessly (left), aircraft-type levers control air conditioning system. Accurate speedometer was a surprise

RUBBER SMOKES during braking test (below)—note how body rolls forward on springs under powerful deceleration. Brake fade was slight

NOTE: In future issues of MOTOR TREND *Griff Borgeson, along with Associate Editor Dick van Osten, will be conducting the majority of our Motor Trials. Griff's many years of automotive background, including several years of field work as a test engineer, give him the objective viewpoint so necessary to the forming of responsible judgments.—Editor.*

IN A WORLD seething with radical changes in automotive design, Lincoln continues with its big, traditional L-head power plant, and with it, against cars powered by admittedly more efficient engines, won this year's Mobilgas Economy Run. Trade scuttlebutt has it that a rival manufacturer with a super-economical design was so certain of winning the Run that thousands of dollars were spent on preparing a publicity campaign based on his make's win. But Lincoln did it. How?

There was the personal factor, of course—the skill of winning driver Les Viland. But what intrigued us most ("us" being MOTOR TREND Research, Dick van Osten and me) was the Economy Run Lincoln's gearing—final drive of 3.31 in conventional, 2.39 in overdrive. Could a heavy car like the Lincoln, geared so high, actually get out of its own way?

The national limelight was on Lincoln, there were questions to be answered, and we called Inglewood Lincoln-Mercury dealer Bob Estes, sponsor of the victorious car. We asked for an exact duplicate of the Economy Run winner, and Bob replied, "I'll go you one better than that. You can have our practice machine—the one we used for test runs to the Grand Canyon."

Frankly, we expected little performance from this car other than good economy at steady speeds. But after almost a thousand miles of driving the machine through traffic, deserts, mountains, at every speed and under almost every road condition, it became apparent that the Lincoln is one of the best cars on the market today, in every way. Here's the story.

Test Report

FUEL CONSUMPTION: Our test car, exact duplicate of the Sweepstakes winner, was equipped with a high-speed rear axle ratio that we'll deal with later, and with .053-in. carburetor jets, which are specified by the factory for cars operating at altitudes of about 5000 feet. Our fuel consumption figures tallied pretty well with the 25.448 mpg average made by Viland's winning car. The average of our own figures for a steady 30 and a steady 45 mph in overdrive was 25.7 mpg. Just to see how much help the force of gravity could give, we took readings on long downgrades. At a steady 60 mph the best we could get was 31 mpg; at a steady 30, 41.5 mpg. This is pretty darned good economy for a 337 cu. in. L-head engine, but if you want to get it when you buy your next Lincoln be sure to specify Grand Canyon Run jets and gearing.

TOP SPEED: This was one of the big surprises of several days of raking the Lincoln over the coals. Les Viland had told us that we could expect to get about 96 mph from the car if we'd really let her unwind. If the carb had been fitted with sea level-specified .055 jets, another five mph or so would have been on tap. Our test strip is about four miles long and within this distance, in spite of the extreme high-speed gear and without over-running our shut-off points, we took the Lincoln through for a fastest run of 100.67 mph, averaged 97.08 over four runs. Even at full throttle there was no perceptible engine vibration and little noise. The car

BORGESON TESTS AMERICA'S ECONOMY CHAMP

simply opens up to full bore, stays there deliberately and happily, decelerates with equal silence and smoothness.

ACCELERATION: The 3.31 "Plains" rear axle ratio is upped to the remarkably close figure of 2.39:1 in OD top gear. There's a widespread suspicion that any car equipped with such a gear can't pull the skin off a rice pudding. Proof of the pudding comes in driving the car—proof that engine and gearing are a match for each other. Further proof can be found in the Table of Performance; the Lincoln's clocked time over the standing quarter was good. Rear axle ratios of 3.91 and 4.27 are optionally available for these cars and will give much livelier acceleration, will make the engine turn over more, use more gas. But in the mountains and in the most dog-eat-dog traffic we found the 3.31 rear end to be fully adequate—in fact, more than equal to most cars on the getaway.

TRANSMISSION: The '51 Lincoln line comes fitted with Hydra-Matic transmission, unless otherwise specified. The Sweepstakes winner was equipped with a standard hand shift gearbox and with Borg-Warner overdrive. This is the same familiar unit which has been winning the public since 1934, has the customary "kickdown" feature for extra steam when you want it. And at speeds below the OD cut-in point—around 25 mph in the Economy Run car—the free-wheeling action of the OD unit permits all gearshifting to be done without use of the clutch—a real convenience in heavy traffic. Engine compression is available for braking throughout the conventional range, and, in OD, above the cut-in point. To get a fuller picture of the current Lincoln line, we ran a few tests on one of the Hydra-Matic jobs. Automatic shifting was a blessing in traffic but our vote goes to the good old quick-acting clutch and the increased economy and control that go with it.

STEERING AND RIDE: We averaged 45 mph over 40 miles of washboard road, and the only vibration noticeable was that which travelled up the steering column. Comfort of the Lincoln is terrific, but you pay a price for it: squishy 8.00 x 15 tires that shriek in agony during even gentle low-speed cornering. This is a typical engineering compromise, a case of not being able to have your cake and eat it too. There's another compromise in the steering, where the popular 5½ turns from lock to lock is used. It's a long way to have to spin the wheel, especially at speed, when split-seconds count. We'd like this car with power steering and about half as many turns required of the wheel. Some American production cars today have rock-steady steering when cornering at high speed, but Lincoln is not one of them. Alertness and correction are required to keep the car in a given groove.

BODY AND INTERIOR: The parts of the Lincoln that meet the public eye are as straightforward and refined as the sound, unspectacular engine-room scene about to be described. One becomes so accustomed to sham vents and meaningless tinsel on current models that at first sight the Lincoln seems plain, almost austere. Our opinion is that the Lincoln is engineered as well as styled, and in good taste. Everything is done properly; the little things are right. Window cranks on the driver's side are laid out so that you don't bark your knuckles on the steering wheel; your arms really relax on the armrests; hands rest comfortably on the doorhandles, and these are designed not to hook clothing.

The interior is spacious, provides excellent comfort for six passengers, is upholstered in rich but restrained nylon and vinyl leather. On freezing desert nights or blazing desert days, the car's ventilating system gave pin-point temperature control. The exterior is handsome and fine, without screaming its price tag to the world. The bumpers are perhaps the most safe and substantial in the field, and chrome is used within nice limits. Fiberglas insulation extends from the front floorboard to top of cowl and over the inside of the entire top, adding a final, nice touch to passenger comfort.

ENGINE: When Viland delivered the Economy Run car to my door I had my pet vintage machine—a 1928 Lincoln touring car—out to meet its newest descendant. The 23-year-old job still goes like a bomb, and Les went over it carefully while I checked myself out on the new car. We talked about the long tradition of Lincoln quality, and about the most significant point of all today, Ford's

CONTINUED ON PAGE 90

BOTH LINCOLN'S visibility and generous luggage space are tops, bumpers extremely rugged

VILAND, BORGESON, and two cars which testify to Lincoln's almost 30 years of V-8 experience

PHOTOS BY E. RICKMAN

'52 LINCOLN

AFTER 3500 GRUELING MILES OF TESTING, HERE ARE THE LITTLE-KNOWN FACTS OF THE ALL-NEW LINCOLN

By Walt Woron

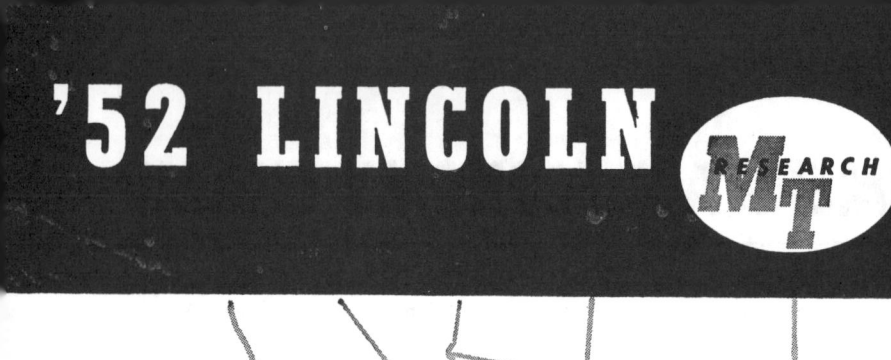

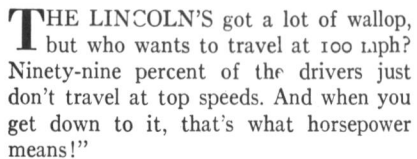

Front bumper and grille are one unit. Vision is unobstructed over low hood

"THE LINCOLN'S got a lot of wallop, but who wants to travel at 100 mph? Ninety-nine percent of the drivers just don't travel at top speeds. And when you get down to it, that's what horsepower means!"

This was the answer to a question we tossed at E. S. MacPherson, Vice-Pres., Engineering, Ford Motor Company, asking him where the so-called "horsepower race" was going to end. He continued, "It's not a sound thing. Passenger tires, particularly, are one weakness—*they're not built for speeds of 100 mph*; highways are another weakness—*they're just not designed for high speed cruising.* You'll find that out on your trip to the West Coast."

And we did. Although we were able to maintain a total driving average of 54.1 mph over a distance of 2486 miles, we had to fight things such as narrow twisting roads, gravel detours, brick-surfaced highways and pock-marked roads.

Handles Better Than Ever

Without a doubt, the most outstanding characteristic of the Lincoln is its handling ability. It doesn't heel excessively on sharp, high-speed turns, and it doesn't feel like you're guiding a couple of sponges around a turn. At one spot we went off the asphalt at a speed of 55 mph, sliding sideways, but it was an easy matter to correct the car's direction. When you slam into a corner hard, you tromp the throttle and it pulls around with no strain. Around sweeping turns at 80 mph neither rear wheel breaks loose, although that is the tendency of most cars. For example, on a left turn, the left rear wheel of the majority of cars tries to lift off the ground (reason, it's the lightest at the time)—not so with the Lincoln.

Steering effort has been greatly reduced, and the suspension action has been improved, both due to the new ball-joint front end design. Instead of using standard kingpins, this design utilizes a ball joint at the wheel location of both the top and bottom A-arms.

Advantages of the ball joint suspension are: easier servicing, more durability, and a reduction in unsprung weight (which results in less steering effort). Lincoln, of course, is not the first manufacturer to introduce the ball-joint suspension, nor is the United States the first country to use it; however, Lincoln *is* the first United States manufacturer to place this system into production.

On the debit side of the handling ledger: Asphalt strips that run parallel to the road direction cause the car to swerve from one side to another, although this is a common fault of large tires; tire squeal at low speeds (which you can blame on the low pressure tires, 8.00 x 15); and the excessive amount of steering effort when parking (also due to the large tires).

Counterbalanced, well-braced hood uses Fiberglas for sound insulation

Gasoline tank filler neck is located behind spring-operated license plate

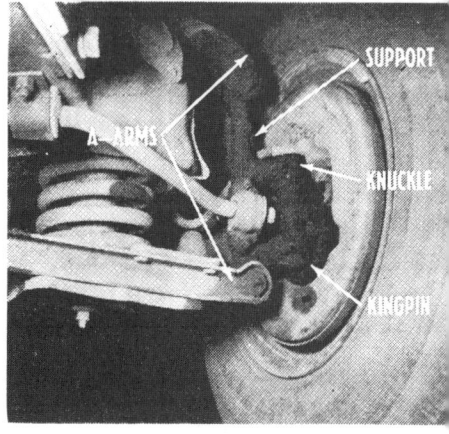

'51 front suspension (above) and '52 (right) shows number of parts used

Battery was placed under front seat board to shield it from engine heat

Spacious trunk has counterbalanced lid that automatically swings open

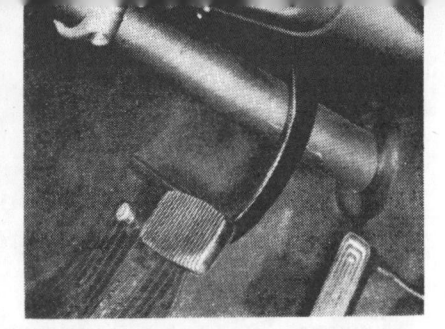

Master cylinder can be served from engine area with pendulum-type brakes

Arizona welcomed us with a snowstorm that made fast driving a hazard

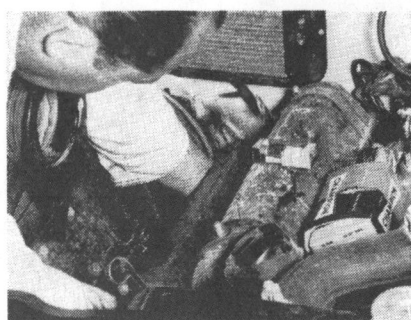

New location of plugs is designed to increase combustion chamber efficiency

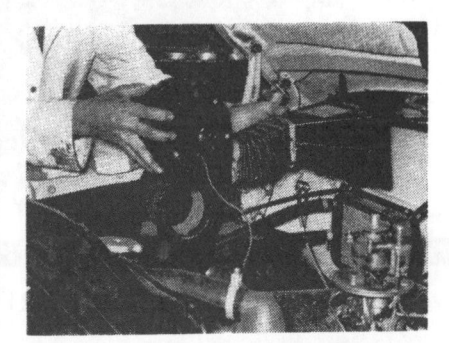

Blower and duct assembly must be removed before getting at spark plugs

You Ride Like a Baby

From the Indiana-Illinois border for a distance of 25 miles we went over a road surface of brick and a brick underbase covered with asphalt that had large chunks of asphalt torn out. This really gave the car a thorough shakedown; bolts that were going to be loosened and rattles that were going to develop would have done so here; but the only discomfort I felt was my concern for the tires. The ride qualities and steering were negligibly affected. During our 3500-mile road test we noticed steering vibration only when driving over asphalt strips in concrete roads.

At speeds of 70 to 80 mph over ocean-wave roads (absolute safe maximum for this type) the Lincoln doesn't bottom, although the suspension is on the soft side and could tend to make persons who are susceptible to car-sickness a bit woozy. The back seat is very comfortable on winding roads at high cruising speeds. Unlike some of the large cars, you don't get whipped from side to side.

Dual-Range H-M Saves the Engine

Since both of the lower-priced cars in the Ford family have their own automatic transmissions, MT Research wondered if Lincoln would be coming out with one of their own design this year. This is a question we asked of Mr. MacPherson. He replied that they would have, except for the lack of machine tools. "In the meanwhile," he continued, "we think the Hydra-Matic is a good transmission, even if it is built by someone else. In my opinion, the Dual-Range Hydra-Matic is good for good drivers; many motorists won't take advantage of it."

What he was referring to in regard to "advantage of it" was the manual selection of gear ratios (third and fourth) while driving in mountains and in traffic. The proper use of this transmission will put money in your pocket, since you can save the brakes and can prevent lugging of the engine. Many persons in H-M equipped cars drive around town using fourth gear most of the time (because of the inherent design.)

Driving properly, you'll use third gear around town—giving you just as much performance flexibility, but with the addition of engine braking—and when you get into the country, you'll shift to fourth, just as you do with an overdrive-equipped car. If you forget to shift to fourth, of course, you'll use more gas. If it's any consolation, an automatic gear shift will make its change around an indicated 85 mph.

The Dual-Range H-M, as with all H-M transmissions tested, has a surge when it changes gears, along with the annoying "clunk" when you drop the selector into reverse. The manufacturers are aware of

Steering is easier through use of ball joints at top and bottom of spindle

CONTINUED ON NEXT PAGE

25

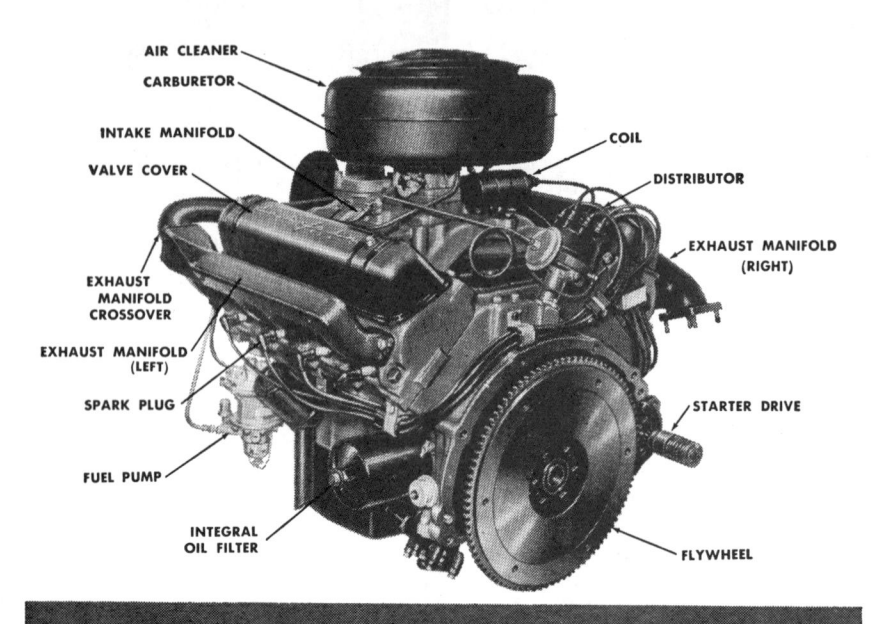

Lincoln's new 160 bhp, overhead-valve V-8 engine gives car plenty of punch

this situation, so it should be corrected in the not-too-distant future.

Brake Pedals Are Pendulum Type

When you step on the brakes you don't actually feel any difference because of the new pedal arrangement—that is, with the pedals hanging down from above instead of coming through the toe board. The brake pedal is wide enough to use a left-foot-brake-and-right-foot-throttle operation if you want. The way the brake and throttle are located there shouldn't be much likelihood of catching the sole of your shoe under the brake pedal as you lift your foot off the throttle.

The brake pedals take more pressure than those of many other cars, but MOTOR TREND Research looks on this as an advantage in that you are not as apt to send the passenger through the windshield by their sudden application.

At no time during the 3500-mile test was there any tendency of the brakes to fade. Stopping distances are about what would be expected of a 4630-lb. car and are about equal to those obtained with the '51 Lincoln, except that at the 45 mph speed of the '52 Lincoln, 14½ more feet of stopping distance were required. This could be due to either the additional weight (225 lbs.) or adjustment.

Acceleration Is About Average

Acceleration qualities of the Lincoln fall in the average range, which is undoubtedly due to the fact that it has a weight loading of 48.7 lbs. per road horsepower and 28.9 lbs. per brake horsepower.

In elapsed times over the standing start quarter mile and also through speeds of 0 to 30 and 0 to 60 mph, the '52 Lincoln with Dual-Range H-M does not stack up to the '51 Lincoln tested (*July '51 MT*) which was equipped with a 3.9 rear axle and standard shift. The higher rear axle of the '51 and the spread in power/weight ratios are undoubtedly the causes of the difference in acceleration qualities. As heavy as the car is, however, it feels like it has more than adequate pickup.

Good Driving Will Get Good Fuel Economy

Fuel consumption of the '52 Lincoln test car was much higher at lower speeds

LINCOLN CAPRI TEST TABLE

PERFORMANCE
CLAYTON CHASSIS DYNAMOMETER TEST
(All tests are made under full load conditions)

RPM	MPH	ROAD HP
1200	33	42
2000	56.5	71
3000	60	(Maximum obtainable) 95

ACCELERATION TRIALS (SECONDS)
Standing start ¼ mile :21.64
0-30 mph :05.18
0-60 mph :16.67
10-60 mph in DRIVE—4th, shifted to 3rd :14.83
30-60 mph in DRIVE—4th, shifted to 2nd :11.40

TOP SPEED (MPH)
Fastest one-way run 100.33
Average of four runs (both directions) 98.31

FUEL CONSUMPTION (MPG)

SPEED	DRIVE—3rd	DRIVE—4th
30	16.3	17.6
45	—	19.7
60	—	16.5
Aproximate average in traffic	13.2	15.6

(Mobilgas Special used throughout entire test)

BRAKE CHECK
Stopping distance at 30 mph 45 ft. 2 in.
Stopping distance at 45 mph 115 ft. 8 in.
Stopping distance at 60 mph 237 ft. 0 in.

SPEEDOMETER CHECK

ACTUAL	INDICATED	ERROR (%)
30	32	6.7
45	50	11.1
60	66	10.0

GENERAL SPECIFICATIONS
ENGINE
Type OHV V-8
Bore and stroke 3.8x3.5
Stroke/bore ratio .92:1
Compression ratio 7.5:1
Displacement 317.5 cu. in.
Advertised bhp 160 @ 3900 rpm
Bhp per cu. in. .504
Maximum torque 284 lbs. ft. @ 1800 rpm

DRIVE SYSTEM
Transmission: Four-speed fully automatic planetary gear set and fluid coupling. Ratios:
First—3.82 Second—2.63
Third—1.45 Fourth—1.0
Reverse—4.3

Rear Axle:
Semi-floating Hotchkiss drive, hypoid gears. Ratios: 3.15:1

DIMENSIONS
Wheelbase 123 in.
Tread 58.5 in.
Wheelbase/tread ratio 2.1:1
Overall width 77.5 in.
Overall length 214.1 in.
Overall height 62.7 in.
Turning radius 22.7 ft.
Weight (test car) 4630 lbs.
Weight/bhp ratio 28.9:1
Weight/road hp ratio 48.7
Weight distribution (front to rear) 55.3 to 44.7

INTERIOR SAFETY CHECK

QUESTION	YES	NO
1. Blind spot at left windshield post at a minimum?		X
2. Vision to right rear satisfactory		X
3. Positive lock to prevent doors from being opened from inside?	X	
4. Does adjustable front seat lock securely in place?	X	
5. Minimum of projections on dashboard face?	X	
6. Is emergency brake an emergency brake and is it accessible to both driver and passenger?		X
7. Are cigarette lighter and ashtray located conveniently for driver?		X
8. Is rear vision mirror positioned so as not to cause blind spot for driver?	X	
TOTAL FOR LINCOLN CAPRI	62.5	

OPERATING COST PER MILE ANALYSIS
1. Cost of gasoline 137.30
2. Cost of insurance 156.50
3. First year's depreciation 585.00
4. Maintenance
   A. Two new tires 61.58
   B. Brake reline 35.10
   C. Major tune-up 8.75
   D. Renew front fender 54.70
   E. Renew rear bumper 34.75
   F. Adjust automatic transmission—change lubricant 11.85
First year cost of operation in cents per mile 10.9¢

—MT Research Staff

Capri Special Custom Coupe

Capri Special Custom Convertible

Capri Special Custom Four-Door Sedan

Lincoln Road Test

than that of the '51 Lincoln. This can be explained by a couple of factors: first, the higher rear axle ratio of the overdrive-equipped '51 (2.71:1 vs. 3.15:1); and second, the '52 test car used an automatic transmission (never as efficient in giving fuel economy). At speeds of 60 mph and in traffic, the two cars are fairly equal in economy.

Actually, the amount of mileage you get while driving in traffic or on the road depends on the way you drive the car. If you storm out at every signal, racing from signal to signal, you're naturally not going to get as good mileage as you would by easing away from a stop, shifting into high gear as soon as possible, anticipating signals and using brakes as sparingly as possible. This has been very graphically demonstrated by the Mobilgas Economy Runs of the past. As a further demonstration of this we tried a fuel consumption test over the same roads, driving in the two above manners. The first time we averaged 9.3 mpg and the second time we got 16.2 mpg. This proves that if you want economy or at least better economy than you are getting, you can get it by careful driving.

Breaks 100 mph One Way

The fact that the Lincoln was able to break 100 mph in one direction, and averaged 98.31 in two-way runs, is an indication that the car has enough horsepower for most people and for most situations. At its top speed and at high cruising speeds the Lincoln feels very safe.

Engine Compartment

MOTOR TREND Research feels that Lincoln has taken a step in the right direction—more efficiency from a smaller engine. Instead of boosting the hp excessively (which they could have done) from the comparatively large (336 cu. in.) L-head engine, they designed a more efficient, smaller (by 18.5 cu. in.), overhead-valve V-8, which still resulted in five more bhp.

Not only has the cubic-inch displacement of the engine been reduced, but the block has been made shorter and weighs 80 to 100 lbs. less. In addition, the piston speed has been reduced by 20 per cent (at the same car speed) over the '51 engine design, which means a longer engine life, since the pistons won't be moving up and down in the cylinder as many times, while the car travels the same distance.

"Why didn't they use the same type of combustion chamber that Chrysler is using—that is, hemispherical—instead of wedge-type?" we asked Mr. MacPherson. His reply was, "Our design is simpler and what we believe to be a better compromise. There might be an advantage to the hemispherical combustion chamber, but we think it's slight."

Power-wise it would appear that the Lincoln is adequate. But in this day of hot competition there is some wonder as to what a manufacturer will have to do to keep pace in the so-called "horsepower race." Undoubtedly, the Lincoln sales people have considered the effect of higher powered, competitive cars on the sales of their own products. And it's a certainty that dealers have voiced their concern over this. It will, therefore, come as no surprise to MT Research to see Lincoln up the hp of their present engine in the not-too-distant future.

A typical comment by service station operators heard while driving cross-country was, "They're sure making these engines handy for us." To some extent this

LINCOLN ACCESSORY PRICE LIST	
ACCESSORY	OVER COUNTER PRICE
Brake signal light	$ 4.95
Seat Covers—Nylon	$ 50.00
Seat Covers—Plastic	$ 44.80
Curb buffer	$22.00 & $ 17.50
Exhaust extension	$ 1.75
Fender shields, pair	$ 21.00
Grille guard	$ 27.50
Heater and defroster	$115.00
Road lamp	$ 15.50
Spot light and bracket	$ 19.50
Back up light, pair	$10.00 & $ 8.00
Outside rear view mirror	$ 5.00
Radio	$104.50 & $ 86.85
Windshield visor	$ 17.95
Rear window wiper	$ 15.10
Windshield washer	$ 7.15
(Note: Prices subject to change without notice)	

is true, since the oil dip stick is easily reached, the H-M transmission's oil level, and the master brake cylinder can be checked from the engine compartment. Unfortunately, there are other points about the engine that make it difficult to service. As an example, the generator is located below the block on the right side and the fuel pump is located below the block on the left side. To get at this you have to come up from underneath. We were told that the generator location was determined by the position of the engine mounts and that the generator runs cooler in this position. The fuel pump was located where it is because it runs 20° F cooler and lifts the fuel about 12 inches less in this location.

Because the engine sump is so low, the lubrication hoist cross-member that fits under the front A-arms could not be used when we serviced the car. Instead we had to place six-by-six blocks on top of the hoist rails to allow the sump to clear.

Interior—Good, Not Plush

The interior of the Lincoln Capri is good without being particularly plush. The upholstery fabric in the test car appeared to be durable and all appointments were well finished. Headroom is good but legroom could be somewhat improved by allowing more adjustment in the front seat track.

Vision is good in all directions, except that the rear view mirror blocks vision to the right front, keeping you from seeing what's coming out of a side street. This could be corrected by a mirror that is adjustable up and down.

All controls are within easy reach of the driver, but the ash-tray could be a little closer.

The hydraulic window lifts (optional on the Capri) are a good feature, but manual window cranks are not as apt to become defective.

Some glare points, all of which are minor, are: the shiny steering wheel, the horn ring, the steering column knob and (from outside) the chrome leaders on the fenders.

What It Will Cost to Run

Taking what we have said about the Lincoln and boiling it down in a few words we can say that it handles very well, has good acceleration, has an exceptional ride, gets fair fuel economy, and style-wise is one of the best. When you try an overall evaluation of *any* car, however, you're handicapped by the fact that so much depends on the wants and tastes of the individual. After all the car's features are thoroughly analyzed and after all the performance figures are studied an important factor still remains, "How much will it cost to keep it running?"

This is where our Operation Cost Per Mile comes in handy. This is a figure based on certain assumed values: that the average car will be driven 10,000 miles a year (the normal average); that the gasoline it uses (regular or premium as required) will cost 25¢ or 27¢; that the insurance carried will be at least minimum coverages (Comprehensive, $50 Deductible Collision, $10-20,000 Public Liability, $5000 Property Damage); that the depreciation will be equal to what it was for the last available period (1950-51) as quoted in the *Kelley Blue Book*; and that during the first year of operation the car will need new tires, brakes, a major tune-up and will have been involved in a couple of minor accidents.

After these assumptions have been made and the costs have been totaled we came up with a figure that indicated what it will cost to run a mile. In the case of the Lincoln, we arrived at a fairly high figure of 10.9¢, largely due to the high depreciation. This has been a shortcoming of the Lincoln for many years, although it is not a surety that the 1952 Lincoln will depreciate as rapidly as its predecessors. Then too, the person who can afford a car in this price bracket ($3729.28 to $4146.38 F.O.B. Detroit) is generally not too concerned with trade-in values. If he gets a good car (as in the case of the 1952 Lincoln) that satisfies him.

—*Walt Woron*

Front end of the new Mercury marks a further step in integrating the bumper into the shape of the car. The bumper has two main bars with an air intake in between. The air intake on the top of the bonnet, like the scoops on the side of the car, is a dummy with no practical purpose at present.

Extensive U.S. Ford Redesign

FORD, MERCURY AND LINCOLN REVISED : NEW O.H.V. SIX- AND EIGHT-CYLINDER ENGINES

INTRODUCTION of the Ford, Mercury and Lincoln ranges for 1952 is an important step in the Ford group's drive to regain the leading place in the United States market which it lost to General Motors some years ago. At a time when their competitors have had to content themselves with face-lifting operations and detail improvements, Ford have completely restyled the whole ranges of Ford, Mercury and Lincoln cars. They have introduced a new overhead-valve six-cylinder engine for the Fords, increased the power of the side-valve Ford and Mercury V8 units and produced an entirely new Lincoln chassis, with an overhead-valve V8 engine and a type of front suspension not previously used in the U.S.A.

An early step, indicating a new attitude to public relations, was when the Ford group, hitherto noted for its reticence, invited the Press to view its new engineering laboratories, as recorded in *The Autocar* of February 1. The long-term experimental car, the Continental Nineteen Fifty X, was shown, and visitors saw many projects now engaging the attention of an engineering and research staff which has been expanded from 1,600 people in 1946 to over 4,000 at the present time. Design and tooling programmes for the new cars were already well advanced before the Korean crisis developed and so the new models have been brought to production despite a growing armament programme.

In style, the Lincoln and Mercury establish distinctive trends which will have considerable influence elsewhere. They are marked by crisp lines and flat surfaces, free from bulbous curves. Radiator grilles are eliminated and bumpers are fully integrated into the styling, instead of being hung on as ugly and cumbersome appendages.

Several mechanical features of the new cars are new to America but have already been seen on European cars. Pedals are now hung from the front bulkhead, bringing the linkage up to where it is less vulnerable, and putting the brake master cylinder where it is more accessible and protected from dirt and ice. The Lincoln has a coil spring and wishbone front suspension in which the separate kingpin is eliminated, ball joints on the ends of the wishbones being used to carry the yoke piece, accommodating both steering and suspension movements. Finally, Lincoln has been able to drop the bonnet line by sinking the carburettor inside the air cleaner, thus cutting the height of the power unit by several inches.

Ford enters the new season with 18 separate models, including two new station wagons, known as Ranch Wagons. They are available in over 80 combinations of colour and trim, and are divided into three main price groups, the Mainline, Customline and Crestline, according to trim and equipment. Bodies have been given a completely new appearance by a restyled grille and bumpers, and one-piece curved screen with narrower pillars plus new doors and rear wings. There is a choice of the new o.h.v. six-cylinder engine or the more powerful side-valve V8, and buyers have the option of three transmissions, a synchromesh gear box, gear box plus overdrive, or torque converter with three-speed planetary gear. A heavier chassis frame is used, with longer wheelbase and wider front track than last year. In the search for lighter steering the steering-box ratio has been changed from 23.2 to 26.3 to 1.

The new six-cylinder engine is known as the Mileage Maker. It has valves inclined in line in wedge-shaped combustion chambers and operated by pushrods. Several features are also common to the new Lincoln V8, including cast-iron alloy crankshaft, camshaft, and exhaust valves. No separate valve guides are used, the guides being cast in the head. The six has nearly equal bore and stroke, the swept volume being 3½ litres. On a compression ratio of 7 to 1

One of the sleekest of the new American models, the 1952 Mercury, avoids bulbous curves. The high rear deck and tail fins characterize the current trend and the big rear lamps merging into the bumper outline set the pattern for future styles.

The hardtop Lincoln Capri, like the new Mercury, has a bumper-cum-air intake which merges smoothly into the lines of the car. Appearance is improved by eliminating the large aircraft or guided missile bonnet emblem seen on so many American designs.

FORD SPECIFICATION

Engine.—6 cyl. 90.4 × 91.4 mm. 3,528 c.c. O.h.v. pushrod. Compression ratio 7 to 1. 101 b.h.p. at 3,500 r.p.m. Max. torque 185 lb ft at 1,300-1,700 r.p.m. V8. 80.9 × 95.25 mm. 3,923 c.c. S.v. Compression ratio 7.2 to 1. 110 b.h.p. at 3,800 r.p.m. Max. torque 196 lb ft at 1,900-2,100 r.p.m.

Transmission.—Standard, single-plate dry clutch, synchromesh gear box. Axle ratio 3.90 to 1, Axle ratio with overdrive 4.1 to 1; with torque converter transmission, 3.31 to 1. Open propeller-shaft. Hypoid axle.

Suspension.—Independent front by coil springs and wishbones, half-elliptic rear.

Wheels, Tyres and Brakes.—6.70-15in tyres in steel disc wheels. Hydraulic duo-servo brakes.

Dimensions.—Wheelbase 9ft 7in. Track (front) 4ft 10in; (rear) 4ft 8in. Overall length 16ft 5½in, width 6ft 1⅜in, height (laden) 5ft 2⅞in. Kerb weight (V8) 3,350 lb approx.

MERCURY SPECIFICATION

Engine.—V8. 80.9 × 101.6 mm. 4,185 c.c. S.v. compression ratio 7.2 to 1. 125 b.h.p. at 3,700 r.p.m. Max. torque 210 lb ft at 1,900-2,200 r.p.m.

Transmission.—Standard, single-plate dry clutch, synchromesh gear box. Ratios 3.73, 5.98, 10.51 to 1. Axle ratio with overdrive, 4.1 to 1; with torque converter transmission, 3.31 to 1. Open propeller-shaft, hypoid axle.

Wheels, Tyres and Brakes.—7.10-15in tyres on steel disc wheels. Hydraulic duo-servo brakes. Braking area 159 sq in.

Dimensions.—Wheelbase 9ft 10in. Track (front) 4ft 10in; (rear) 4ft 8in. Overall length 16ft 10in, width 6ft 1½in, height 5ft 2½in.

LINCOLN SPECIFICATION

Engine.—V8. 96.5 × 88.9 mm. 5,202.7 c.c. O.h.v. pushrod. Compression ratio 7.5 to 1. 160 b.h.p. at 3,900 r.p.m. Max. torque 284 lb ft at 1,600-1,900 r.p.m.

Transmission.—Hydramatic 4-speed epicyclic with fluid coupling. Axle ratio 3.15 to 1.

Suspension.—Coil springs and wishbones with ball joints for yoke piece at front. Half-elliptic rear.

Wheels, Tyres and Brakes.—8.00-15 on steel disc wheels. Hydraulic brakes, 11in drums. 191 sq in braking area.

Dimensions.—Wheelbase 10ft 3in. Track (front and rear) 4ft 10½in. Overall length 17ft 10½in, width 6ft 5⅜in, height 5ft 2⅛in. Kerb weight (4-door saloon) 4,060 lb.

it delivers 101 b.h.p. with a maximum torque of 185 lb ft. The classic side-valve Ford V8 engine is continued, with compression raised from 6.8 to 7.2 to 1 and it now gives 110 b.h.p. as against 100 last year.

The Mercury, apart from new styling, has more body space than before and the chassis is reinforced, with box-section side members instead of the previous channels. The range of body styles is the biggest ever offered by Mercury, with hardtop coupés and convertibles, plus two- and four-door station wagons supplementing the two- and four-door saloons. The facia designs are interesting, with instruments centrally grouped in one big dial and control levers arranged vertically as on a modern aircraft. The windows are larger than before and there is more head room. Steering effort has been reduced by giving further reduction in the box, the ratio now being 26.4 to 1. A rise in compression ratio similar to that on the Ford V8 has raised the Mercury power output from 112 to 125 b.h.p.

The Lincolns look remarkably long and low, but they have higher doors and more head room than the previous models and there is more leg room at the rear. Despite all this, the cars are 8in shorter and 1in narrower than the 1951 models. The new chassis frame has a massive centre X bracing with additional lateral reinforcements. The low short bonnet should ensure good driving vision and Lincoln now conforms with convention by hinging all doors on the forward edge for safety. Reduction of weight has made it possible to reverse the present trend, by using a lower reduction in the steering. The ratio is now 26.1 to 1 instead of 29.7 on the previous cars, so Lincoln owners should have less wheel twirling to do.

The battery is now located under the front toe board, where it can be reached from inside the car, and is of a new type, with 63 very thin plates inside a moulded-rubber case, instead of the previous seventeen. It is said to give a longer life, with less maintenance than before.

When it comes to making V8 engines the Ford group has, of course, unrivalled experience, for they have built over twelve million of them, which they believe is

The tail of the Lincoln Cosmopolitan saloon exhibits the high flat line and enormous tail lamps which seem destined to appear in the next series of American cars. Even on the saloons the rear quarter panel has practically disappeared.

29

New grille, lamps, bumpers, rear doors and rear wings, plus a curved screen, change the appearance of the 1952 Detroit Fords, which are available with a more powerful side-valve V8 engine or a new o.h.v. six. The air intake on the side is a dummy with no functional significance.

Extensive U.S. Ford Redesign
continued

The new o.h.v. Lincoln V8 engine is given an unusual appearance by offset rocker boxes and exhaust manifolds rising from the centre of the cylinder heads. The carburettor is sunk inside the air cleaner to permit a lower bonnet line.

On the Mercury all instruments are grouped in one central dial and an idea from aircraft practice is borrowed by using vertical levers to control the ventilation system.

more than four times as many as all other manufacturers put together. The new engine is smaller and lighter than the unit which it replaces but gives more power and a higher torque. Its output is 160 b.h.p. from a swept volume of 5,202 c.c. whereas the previous engine gave 154 b.h.p. from 5,517 c.c. The design of the new Lincoln V8 is, therefore, of special interest. The cylinder blocks are set at 90 degrees in a single casting with a crankcase which extends down well below the centre line of the crankshaft, unlike some other recent designs. The valves are inclined in wedge-shaped combustion chambers, the Lincoln designers having decided against the hemispherical head on the ground of complication and tooling difficulties. The valve cover is offset to one side of the head and the engine is given an unusual appearance by the exhaust manifolds, which rise straight from the centres of the cylinder heads. Manifolds are finished with silicone heat-resisting paint. The automatic choke is mounted on the inlet manifold, a location adopted on last year's Detroit Fords. This causes the choke to operate in accordance with engine manifold temperature rather than carburettor temperature.

Crankcase ventilation is accomplished through an intake which allows air to flow through the tappet chamber, chain cover and rocker boxes and then into the crankcase at front and rear, after which it is finally ejected through exhaust ducts under the car. Removable filters exclude grit and dust. It is claimed that the air flow is automatic and does not depend on the movement of the car or the action of the fan. The fuel pump is placed low down towards the front of the engine in the path of cool air from the fan.

The Cosmopolitan and Capri series replace last year's Cosmopolitan Lincolns, the short-wheelbase series being dropped. All cars are equipped with the Hydra-matic transmission.

Sales to dollar buyers in Britain are handled through Lincoln Cars, Ltd. Great West Road, Brentford, Middlesex.

Out on a limb—perhaps, but our expert test driver says—driving the '52 Lincoln was the nearest sensation to flying—it's a Pullman of the highway—tops in roadability, ease of handling and overall driving characteristics.

Speed Age Tests The 1952 Lincoln

Walt Woestman Photo

Encountering nearly every condition under which an automobile could be expected to operate, the 1952 Lincoln performed each test with confident ease and ability. Spectacular acceleration gave test expert Ted Koopman much driving satisfaction.

By TED KOOPMAN
ASSOCIATE EDITOR, SPEED AGE

NEW FROM heel to toe, the 1952 Lincoln represents nearly four years of exhaustive research by engineers whole sole aim was the development of an overhead valve, high efficiency V8 engine and a chassis capable of utilizing the enormous power output latent in this power plant.

Incorporating a myriad of engineering innovations, months of intensive study and trial were required to make each phase conform with the exacting standards and tolerances essential to precision engine production.

Costly though it was, Lincoln officials are confident the extended trial and test period has produced a car essentially free from the mechanical difficulties usually encountered in first models, and the next few months will determine whether their supposition is correct.

Without question, this 1952 Lincoln is a car of unusual merit and I find there aren't enough adjectives to properly describe its exceptional performance during the 2,000 miles I covered.

However, I am satisfied no other production car can equal its roadability, ease of handling and excellent overall driving characteristics, and though I realize the magnitude of this statement, actual performance during this road test leaves me no alternative.

Its spectacular acceleration and top speed give driving satisfaction difficult to compare and the weight distribution, suspension and superior shock absorber control make boulevards of plowed fields, all without the sacrifice of operating economy.

Entirely new in design, the Lincoln's backbone is a rigid box channel frame, with an I beam, X brace and five cross braces at critical points, assuring a distortionless unit under all road conditions.

Similar to the Jaguar's, but of fortified design, the front suspension employs two ball and socket joints on each wheel; one connects to the top supporting arm, the other the bottom.

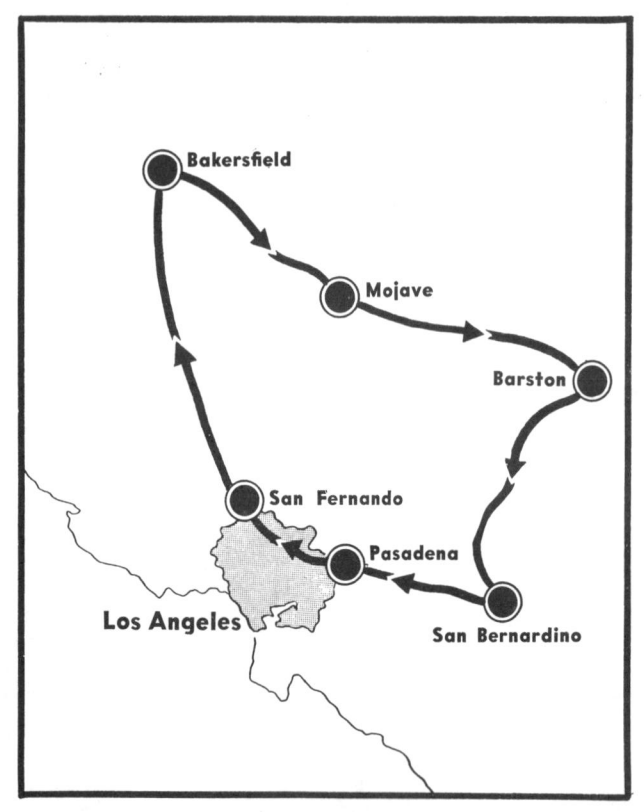

A well planned road test uses the varied conditions that motorists would be likely to encounter in traveling. Route of the Lincoln took 2,000 miles of highways for testing purposes.

The 1952 Lincoln supplied for testing was from showroom of Bob Estes, Inglewood, Calif., Lincoln-Mercury dealer. Estes is well known in sprint and championship competition and his Offy was slated to be driven by Jim Rigsby in the '500'.

Walt Woestman Photo

This eliminates the king pin support forgings, allowing vertical motion to absorb road shock and rotary movement for the easiest mechanically controlled steering available.

Its incredible directional stability and low center of gravity give glue-to-the-road handling characteristics far beyond the working limits of other styles, without the front wheel scuffing or unpleasant body sway common to the majority.

Rear suspension is by two longitudinal semi-elliptic springs whose rubber bottoming bumpers have been designed to give increased dampening in direct proportion to the tension developed by the leaves, thereby easing and quieting the impact when bottoming does occur. Double acting shocks are mounted at a diagonal to minimize kick-up or excess vertical motion.

The steering assembly, sturdier than the 1951 design, has improved connections to the pitman arm, and needle bearings replace the usual bronze sleeve bushings. The brake pedal is

Rugged backbone of the Lincoln is a rigid box channel frame, with an I beam, X brace and five cross braces at critical points. Chassis assures distortionless unit under all conditions. Front suspension uses two ball and socket joints on each wheel.

Dynamometer tests were ably conducted at Paul Brothers, Inc., Washington, D. C. under the direction of shop manager John Major, left, and assisted by mechanic Roger Burriss.
Dick Adams Photo

The power plant of Lincoln is a completely new high compression V8 engine that incorporates overhead valves and rocker arms that are actuated with valve rotating hydraulic lifters.

supported from an assembly behind the steering column, rather than under the floorboard, allowing installation of the master cylinder on the forward side of the fire wall with simplified linkage that reduces braking pressure requirements.

The outstanding development, however, is the new high compression V8 engine, incorporating overhead valves, rocker arms actuated with valve rotating hydraulic lifters.

The cam lobes have been reshaped for smoother and quieter action and valve guides bored directly in the block casting reduce exhaust valve temperatures approximately 100 degrees.

Although smaller than the '51 model by 20 cubic inches, this engine develops more horsepower due to the better breathing provided by use of larger intake valves, non-restricted intake manifold and improved carburetion. The automatic choke, located on the intake manifold, operates in accordance with engine manifold temperatures rather than those of the carburetor throat and provides a more accurate control of overchoking and its accompanying evils.

The extended crankcase is an exclusive Lincoln feature and by carrying it below the center line of crankshaft, rigidity is increased, a better mounting is provided for the fly wheel housing and a wider gasket surface produces less oil leaks between the case and oil pan.

Another exclusive Lincoln feature is the method of anchoring head bolts in the block, eliminating head distortion due to uneven cylinder head tightening.

A higher cranking ratio of the starting motor to flywheel and improved spark control provide quicker starts in cold weather and the more rapid ejection of exhaust gases made possible a reduction in the cooling system capacity from 34½

Walt Woestman Photo

Instrument panel is simple and effective. Position of controls are within normal reach. Brake is readily accessible and the positive ratchet eliminates chance of accidental release. Two-spoke steering wheel reduces chance of chest injury.

Walt Woestman Photo

Major John C. Giraudo, on wing, Speed Age's contest winner for best letter on auto test report, and wife, feminine test driver; both were more than satisfied with power and handling ability of Lincoln car after extensive tests in Nevada.

to 24½ quarts. This cut in radiator dimensions conserves copper and the smaller fan requires less horsepower to drive.

Although the new exhaust system gives greater gas rejection, dual pipes would be even more efficient. The present use of a muffler and resonator in the single exhaust line seems contrary to all established practice for maintaining a low back pressure. A new type battery, smaller and 10 pounds lighter, has 63 thin plates and is located under the right front foot board. This may be an ideal location for protection of the battery but since it is inconvenient to get at, I wonder if it will receive normal servicing.

The discontinuance of syncro-mesh transmission and overdrive as optional equipment will be a disappointment to some, but the dual range Hydra-Matic operating with a 3.08 pinion gearing, provides the economy and quiet cruising characteristics of the overdrive while the fourth speed lock-out gives flexible traffic operation combined with exceptional acceleration.

From the view point of the mechanic, all is not so rosy and the location of the oil filter promises facial oilings for unwary changers of cartridges. The changing of spark plugs while exhaust manifolds are hot will be another interesting detail of service.

The interior of the body is impressive in its simple luxury with the finest materials beautifully styled and executed. Seats are deep and backs provide comfortable support with leg room ample for completely relaxed riding positions. All body hardware is of highest quality and lighting fixtures, switches and ash trays are conveniently located.

The increased area of glass surface provides extensive vision in all directions, but at the same time gives the impression of riding in a gold fish bowl, with an equal amount of privacy.

I am not convinced that good vision requires so much glass, and have found it extremely unpleasant for travel in Southern California, Nevada, Arizona and other plain and desert states having a high range of temperature. I don't refer to the Lincoln alone, rather to any car having excessive glass, for under those conditions, it becomes a definite safety hazard, with the intense heat and concentration of ultra violet rays creating a toxic, sleep-inducing atmosphere.

The addition of side awnings, sun visor and venetian blind in the rear window makes it somewhat more bearable, but still decidedly uncomfortable.

The body design appears to be highly controversial and its approval is somewhat dependent on the manufacturer's aims, for if it was intended to be simply another automobile, the result is adequate. But if the primary objective of the Lincoln Division was to produce a car of distinction comparable with their Lincoln Continental—individuality—the fundamental requirement is obviously missing.

When Lincoln discontinued the Continental in 1948, they relinquished their position in the class car inner circle, leaving Cadillac in sole possession of that coveted honor. The Cosmopolitan of 1949 was undoubtedly intended to occupy the spot vacated by the Continental, but failed dismally to meet that standard, and while improved each year, it was unable to earn the respect of discriminating purchasers whose approval is a basic requirement for class recognition.

Chrysler made their bid with the 180 Imperial, and though a credit to its maker, failed to crash

Ted Koopman Photo

Walt Woestman, six-foot, six-inch associate editor of Speed Age, illustrates exceptional large trunk compartment of '52 Lincoln. Walt accompanied Koopman on road test journey

TESTING THE LINCOLN

the barrier surrounding that enviable position.

With the year's outstanding engine and chassis, Lincoln might have been in scoring position, and while this 1952 model will acquire prominence and earn the respect of all concerned, the body lacks the distinguishing element necessary to regain the position vacated by the Continental.

Actually there is little original design incorporated in this body and its appearance might be said to bear mute testimony to the excellence of competitor's design.

The tail light motif is effective, but an original design would have personalized rather than confused the car with Cadillac, and the adoption of Oldsmobile's simulated rear fender air scoop, familiar to all as a Cadillac-Oldsmobile innovation, hardly fits the advertising matter which declares: "A distinctive (?) side-scoop styling accentuates the sweeping lines of the car."

True, the lines start to sweep from front to rear, but rather than accentuating, this air scoop chops them off abruptly, creating the impression of stubbiness. In time of war, the Navy paints ships with similar diagonals of varying size and angles to distort the profile and confuse the enemy and the Lincoln camouflage has accomplished the same result.

From the utility standpoint, the body offers the utmost in convenience and comfort, with doors hinged at the front for safety and accessibility. The trunk compartment is exceptionally large as the relaxed position (see photograph) of my co-pilot and photographer, six foot, six inch 'Colonel' Walt Woestman readily demonstrates.

The new type trunk lock is convenient, but lacking a handle, opening may be difficult when it is covered with snow and ice. The same criticism applies to the gas filler cap located behind the number plate.

I was fortunate to have no real need of the bumper jack supplied with the car, but curiosity prompted me to experiment with the two-legged affair that permits operation without involving calisthenics, and if anything could make tire changing

Standard Specifications and Performance Data 1952 Lincoln

Width of front seat measured
5 inches from back .... 62.6 inches
Width of back set measured
5 inches from back .... 62.1 inches
Depth of front seat cushion .. 18.5 inches
Depth of rear seat cushion .. 18.9 inches
Height of front seat cushion .. 12.9 inches
Height of rear seat cushion ... 4.1 inches
Front seat horizontal adjustment .... 12.5 inches
Vertical distance wheel to seat .... 5.8 inches
Head room front seat .... 36.1 inches
Head room back seat .... 34.70 inches
Leg room front seat .... 43.1 inches
Leg room rear seat .... 41.62 inches

Engine Specifications

ENGINE:
Number of cylinders 8
Arrangement .... 90° V OH Valve
Bore .... 3.800 inches
Stroke .... 3.5 inches
Displacement .... 317.5 cubic inches
Taxable horsepower 46.2
Brake horsepower .. 160 HP @ 3900 RPM
Maximum torque 284 ft. lbs. @ 1800 RPM
Compression ratio .... 7.5 to 1

PISTONS:
Make .... Own
Material .... Aluminum alloy
Features .... Slipper type skirt
Weight .... 22 ounces
Compression rings 2
Oil rings 1

CONNECTING RODS:
Length C. to C. .... 7.064 inches
Weight .... 28.22, less bearing

CRANKSHAFT:
Material .... Cast alloy iron
Weight .... 66 pounds
Number main bearings 5
Connecting rod journal
diameter .... 2.2486 inches

VALVE TIMING:
Intake opens .... 18° BTC
Intake closes .... 58° ABC
Exhaust opens .... 56° BBC
Exhaust closes .... 20° ATC
Firing order 1-5-4-8-6-3-7-2

MISCELLANEOUS:
Oil capacity (including filter) .6 quarts
Gas capacity .... 21 gallons
Water capacity, with heater 24.5 quarts
Spark plugs .... 14 MM

COMPRESSION PRESSURE at
cranking speed .... 120 @ 120 RPM

Chassis

FRAME:
Type Closed channel rail ...... X brace
WHEELBASE: .... 123 inches
TREAD:
Front .... 58.5 inches
Rear .... 58.5 inches
WEIGHT:
Curb .... 4060 pounds
DIMENSIONS:
Overall length .... 214.1 inches
Overall width .... 77.5 inches
Overall height .... 62.65 inches
REAR AXLE:
Type .... Semi-floating
Gearing .... Hypoid
Drive .... Hotchkiss
TRANSMISSION:
SYNCHRO-MESH .... Not available
HYDRA-MATIC DUAL RANGE:
Number forward speeds 4
Down-shift possible up to .. 60 MPH
Ratios: first .... 3.8195 to 1
second 2.6341
third 1.450
fourth 1.000
reverse 4.3045
FOOT BRAKES:
Drum diameter .... 11 inches
Material ..... Composite pressed steel and cast iron
Effective area ... 202.34 square inches
Percent effective rear 41%
Type .... Duo Servo, single anchor
Linings .... Moulded asbestos

STEERING:
Type ..... Worm and triple tooth roller
Wheel diameter .... 18 inches
Turning radius .... 24.33 feet
ROAD CLEARANCE:
Minimum .. 7.2 inches at No. 2 cross brace
At rear axle .... 8.2 inches
TIRES:
Standard .. 8.00 x 15 super balloon, 4 ply
Optional .. 8.00 x 15 super balloon, 6 ply

Performance Data
Rate of Acceleration

0 to 30, low only .... 4.15 seconds
0 to 60, OD locked out .. 13.57 seconds
20 to 50, down shift ...... 9.2 seconds
30 to 50 .... 7.61 seconds
30 to 60 .... 10.05 seconds
40 to 60 .... 7.72 seconds

Top Speed   Corrected

Average of four runs .... 102 MPH
In opposite direction .... 99 MPH
Actual top speed .... 100.5 MPH

Speedometer Correction

Indicated   Actual
20 MPH .... 17 MPH
30 MPH .... 28½ MPH
40 MPH .... 39 MPH
50 MPH .... 50 MPH
From 50 MPH to 100 MPH, no error.

Dynamometer Test—Full Load

1200 RPM—31 MPH .. 52 Rear axle horsepower
2000 RPM—58 MPH .. 78 Rear axle horsepower
3000 RPM—88 MPH ... 104 Maximum R.A. HP.

Fuel Consumption

Steady 30 MPH .... 21.5 MPG
Steady 40 MPH .... 21.7 MPG
Steady 50 MPH .... 17.8 MPG
Steady 60 MPH .... 16.4 MPG
Average 497 miles .... 17.2 MPG

Other Cars Road-Tested by Speed Age

Hudson Hornet—Motor Survey Issue
Nash Rambler—January 1952
1952 Dodge—February 1952
1952 Pontiac—March 1952
1952 Nash—April 1952
1952 Mercury—May 1952

Coming Next Month: New Aero Willys and Packard

TESTING THE LINCOLN

more bearable, that jack would lead the parade.

Material shortages continued to delay Los Angeles production of Lincolns, and finally, through the generosity of Bob Estes, Inglewood Lincoln-Mercury dealer well known in sprint and championship competition, a car was made available for testing, but carrying dealer's plates, its use was restricted to California.

Consequently, the trip through Death Valley and Las Vegas was abandoned and a shorter, but equally demanding route was laid out by Walt Woestman, and with sufficient photographic equipment to stock the average camera shop, we rolled down his driveway in Altadena, headed for Bakersfield.

The front compartment of the Lincoln is unusually roomy and the position of seats affords the utmost comfort whether driving or a passenger. Just watching Walt stretch those number 13 brogans and six-foot, six-inch frame, while remarking, "And room to grow, too," was evidence of its exceptional dimensions.

Although the road north is designed for fast travel, we took it easy most of the way, holding an average of 60 miles an hour up hill and down and resisting the temptation to tramp on it when the five mile and longer straight sections of highway stretched to the horizon.

The car showed no desire to 'hunt,' and steering was so effortless it seemed as though we should be able to set a 'trim-tab' and give the car its head. In the mountains where canyons are numerous, wind turbulence is often violent, yet only the strongest gusts created any deviation from a straight line of travel, and then but slightly.

Travel in this car is the nearest sensation to flying possible and the term 'pullman of the highway' would be no exageration.

We arrived in Bakersfield at lunch time, and on refilling the tank, found our average had been approximately 16 miles per gallon for that leg. Our immediate destination was Barstow, via Mojave, on Route No. 466, where we encountered nearly every condition under which an automobile could be expected to operate. Finally, a long stretch of smooth, straight road broke down my resistance and I gradually brought the needle up to the 100 MPH mark and five more, for good measure, with the Tapley performance meter still showing acceleration I backed off.

However, a tail wind and one-half percent down grade is hardly a test for top speed, but it was of considerable value in determining the car's high speed handling characteristics, which actually were no different than at 60.

Three hours out of Bakersfield, we pulled into Barstow, and while too early for supper, did partake of what Walt declared to be tea and crumpets, an affectation, no doubt, from his association with the 'Stuffed Shirts' (April 1952, SPEED AGE).

On the final leg between Barstow and Victorville, the road is a continuous series of dips, some of which will give a car quite a tossing around if traveling in the upper sixties or seventies, but the spring action and control of the Lincoln permitted speeds in the eighties without difficulty and a person riding along on the rear seat would have suffered no discomfort other than the normal bouncing.

Rolling down the San Bernardino side of the Cajon pass at speeds best not declared but well over established limits, the stability of the front suspension was conclusively demonstrated and it seemed as though we were running on rails.

His background of track and highway driving demands that a car must give an unusual performance before Walt displays any enthusiasm and he agreed the Lincoln's cornering and general roadability was the best he'd experienced, outside of cars with specially designed equipment.

Our average gas consumption for the trip, while only approximate, was slightly over 17 miles per gallon and considering the type of country traversed and my total disregard of economical driving, I consider it a remarkable showing.

Having had considerable correspondence with Major John C. Giraudo stationed at Nellis Field, Nevada, Winner of the SPEED AGE contest for best auto test report letter, February issue, I asked him to test drive the new Lincoln and let me have his reaction.

Unable to take the car I was using across the line into Nevada, I was obliged to make arrangements to pick up a car registered in Nevada, and after listening to my tale of woe, the Las Vegas Lincoln-Mercury agency, placed a '52 model at my disposal for further tests.

With Major and Mrs. Giraudo sharing the driving, we made a number of sightseeing trips through that most interesting countryside. (Ed Note: We had to assure Air Force officers that the Major's statements are his own and do not constitute an official endorsement of the Lincoln or SPEED AGE, typical of the Government red tape.)

In Major Giraudo's words: "I was honestly surprised that a Lincoln was so much better than other cars I've driven recently. My recollection of previous Lincolns was of the lack of power in the 12-cylinder cars and clumsiness in the later eight-cylinder jobs.

"This 1952 Lincoln has neither fault, nor any particular fault as far as I am able to determine, except that I do think the excess road clearance of the body detracts from the car's appearance. I would also prefer less miscellaneous chrome, especially where it serves no purpose.

"Might I suggest you offer one as a first prize in your next contest."

Mrs. Giraudo's reaction was similar, as her remarks disclose. She said, "The car seemed so large it frightened me, but in a short time I lost all fear of driving it and discovered it was easier to handle than some smaller cars I've driven.

"The seat made it easy to see the road and the steering wheel is in just the right position. Being able to see both front fenders is very assuring and a big help in driving through traffic, and although it appears to be a man's car, it has all the qualities a woman enjoys in an automobile."

The driver and passenger safety potential is somewhat above average in all Lincoln bodies. Driving visibility is extremely good and although the roof supporting posts are somewhat heavy, their position away from the front corner, made possible by the curved windshield, offers little obstruction to forward vision.

The two-spoke steering wheel reduces chance of chest or stomach injury so prevalant with three-spoke wheels, and the recessed glove compartment knob reduces possibility of head injury showing the growing trend among manufacturers to give greater consideration to rider safety.

The emergency brake is readily accessible and the effective and the positive ratchet eliminates chances of accidental release. The high rate of acceleration coupled with efficient brakes and superior directional change stability provide a maximum possibility of avoiding danger.

Outside the usual destructive hood ornament, the exterior is neither exceptionally safe or outstandingly dangerous and must be accepted as more or less representative of American production cars.

In the 1952 line, the Cosmopolitan is the lower priced model with the Capri being the deluxe style. Each series offers a four and two-door sedan, a sport and hard top coupe, plus a soft top convertible in the Capri series only.

During the period covered by this test, I drove four Linclons, three with regular Hydra-Matic and one with the Dual Range, but unfortunately was unable to try one with syncro-mesh transmission and overdrive.

The Dual Range Hydra gave a much better performance and inasmuch as it will be the only transmission available, all performance tests were made with that type transmission.

Although a bit early for soup-up equipment to hit the market, this car, with its five main bearing crank and short stroke would appear to be a natural for souping, and I believe interesting developments along that line are due this year. ☆ ☆

NEW CARS DESCRIBED

Power steering, power brakes and a power-operated four-way front seat adjustment are optional equipment on Lincoln models for 1953. Lincoln is in the forefront of the current American horse-power "race" with an output of 205 h.p. from the V-eight engine. This is the Capri hard-top coupé.

LINCOLNS WITH 205 b.h.p.

MODIFIED ENGINES AND EXTENSIVE AUTOMATIC EQUIPMENT FOR 1953

WORLD-WIDE attention has been focused on the 1953 Lincolns by the remarkable feat of their taking first four places in the production car division of the Pan-American road race in Mexico, with an average speed of 90.9 m.p.h. by the first car for the whole of the 1,934 miles.

The power output of the o.h.v. V-eight has been raised to 205 b.h.p., making Lincoln the first manufacturer in the world to offer more than 200 h.p. in a standard quantity-production saloon.

The engine, which has a swept volume of 5,204 c.c., now operates on a compression ratio of 8 to 1 instead of the former 7.5 to 1 and has a new four-choke carburettor. At low speeds, only two chokes are in operation, but when additional acceleration is required the two additional chokes are brought into operation automatically. The intake manifold has been redesigned to suit; inlet valves are larger, and a modified kidney-shaped combustion chamber has been introduced to promote turbulence. The engine's breathing capacity has been increased by the use of a new air cleaner and by an improved exhaust system, said to reduce back pressure by anything up to 50 per cent.

European Feature

Lincolns have a simple ball joint steering and suspension assembly of a type common in Europe but hitherto unknown in America. The combined yoke-piece and king pin is carried directly in ball joints on the ends of the wishbones, replacing the multiple joints of the conventional yoke-piece and steering swivel assembly. For 1953 an hydraulically operated power steering unit is optional, reducing the effort required at the steering wheel rim when parking the car from 32-45 lb to 4-7 lb.

Vacuum servo brakes with a vacuum reservoir tank are another optional extra on the latest models, reducing the pedal pressure required by 30-40 per cent. A further innovation is an automatically adjustable front seat in which a series of electric motors operated by push buttons at the driver's side permit adjustment both vertically and horizontally. No wonder Lincoln publicity exults: "Lincoln is the most powered automobile in production today"!

Styling changes are slight, the general appearance closely following the new lines introduced early this year, but the lingering doubts which so often seem to assail American manufacturers as to the ability of the public to distinguish their car from its competitors are reflected in the use of block letters for the name Lincoln on the front. All models now have a single-piece curved rear window, and small gold V motifs have been introduced in the styling of the front grille assembly and on the side panels.

Modified Lincoln Engine in American-owned Allard

EIGHT carburettors (four double-choke) are used on the modified Lincoln V-eight engine used to power the Allard, as seen in the photograph. The modifications are those of Ed. Winfield, of Los Angeles, U.S.A., and include boring out from the standard (1952) 317.5 cu in (5,204 c.c.) to 352 cu in (5,769 c.c.). Pistons are now 4in in diameter. The camshaft has been reground and the hydraulic tappets are eliminated. The unit peaks at 6,200 r.p.m.

The car itself has been modified. The front axle is a Ford truck pattern, divided and hinged, and the transmission is standard Ford with Zephyr gears and a Ford differential with Cyclone assembly permitting 30 different gear ratios. The rear end is de Dion.

The Lockheed brakes have Al-Fin drums, and the car is, as a whole, built to Sports Car Club of America class 1 regulations; it will be entered in Californian sports car races in 1953, where it is sure to be watched with considerable interest.

The compact V-eight unit fits well back in the Allard frame; four long branches feed into an exhaust pipe on either side.

'53 Lincoln Preview

Here's a car with power seating—cushions that rise one and a half inches or move horizontally by the flick of a button.

SPEED AGE STAFF REPORT

Photos by DON O'REILLY

THE Lincoln for 1953 was introduced on November 25 in dealer showrooms across the nation but a press preview a few weeks before, gave SPEED AGE staff members an opportunity to drive the powerful model over Ford's Detroit proving grounds.

Perhaps the most astounding news, however, was that, in a secret series of speed tests with the '53 engine in a '52 chassis over Utah's Bonneville Salt Flats in August, this car had turned 117 MPH.

AAA timing equipment recorded each test sponsored by the factory and Bill Stroppe, who won the 1952 Mobilgas Economy Run in a Mercury, took a Capri through the traps. This amazing speed unofficially cracked more than 30 American stock car records.

The press preview included a visit to the factory showroom where several '53 models were unveiled. Following several discussions by technicians and engineers at the factory, the party moved to the proving grounds for comparison tests of the '52 and '53 models.

Speed Age Staff Writer Bob Russo behind the wheel of the new '53 Lincoln at Ford's Detroit proving grounds.

Last year's Lincoln fitted with the new 205 HP engine for test purposes. The new Lincoln retains the massive dual bars that formed the basis for the grille of the 1952 models. The medallion has been dropped and lettering changed on the '53.

The first, for acceleration, was viewed from atop the 50-foot hill used for load and pull value. Below, on a straight and level runway, a '52 Capri was lined up with a '53 model.

At the drop of a flag both drivers accelerated, with the '53 jumping ahead. Before they reached the 500-foot marker, the new model was two car lengths in the lead and at the 1500-foot line, the '52 was completely out of the picture.

A trip around the huge, banked track followed the match race. SPEED AGE staff members were treated to a 100-MPH-plus ride on the paved circuit. It was a smooth trip and except for the centrifugal force causing the car to lean into the high banks, might have been only a Sunday afternoon drive.

After a luncheon at the Dearborn House, the '53 was placed at the disposal of SPEED AGE for tests.

According to factory technicians, the linking of distributor and carburetor for immediate spark advance makes for faster and smoother acceleration.

This was borne out by test results shown below:

From 0 to 60 MPH, 12 seconds in the '53; 16 seconds in last year's model with a 0 to 80 MPH figure of 22 seconds as compared to 31 seconds.

On the pull and load test, the Lincoln easily pulled its four heavy occupants up a 50-foot incline from a standing start, the Hydra-Matic transmission automatically shifting to a lower gear when the gas pedal was completely depressed.

Power steering, incorporated by Lincoln, results in smooth, effortless handling and a very definite difference can be felt when returning to the conventional steering method after a few miles behind the wheel of the new model.

Parked on dry concrete, the Lincoln equipped with power steering requires only four to seven pounds effort to turn the front wheels as contrasted to 32 to 45 pounds needed for the conventional type. With a Lincoln moving at 10 MPH, the steering effort is only four to five pounds.

Powered by the new engine, a 1952 Lincoln tops a terrifically steep incline at the Ford Detroit proving ground.

The 1952 model begins its descent of the man-made hill. The new model continues the fender-embracing bumper.

39

The '53 features a 205-horsepower engine, the most powerful ever offered in a modern production automobile. The Cosmopolitan and Capri are equipped with power brakes and a 4-way front seat adjustment which operates under power. Both are optional.

The power brakes are new with Lincoln and require 30 to 40% less pedal pressure. The accelerator and brake pedal are located approximately the same height from the floor, allowing a quick swing from one to the other with a minimum effort. A vacuum tank gives added protection for emergency use in the event of power failure.

The power seat is ideal for men and women who have difficulty in adjusting themselves to the correct driving position. It permits an infinite number of seat adjustments with a maximum of one and a half inches vertical and four inches horizontal movement by means of electric motors activated by push buttons on the driver's side of the seat.

An advanced design 4-barrel carburetor, a redesigned intake manifold and larger intake valves combined with a new combustion chamber, a straight-through exhaust system and an improved air cleaner serve to give added power at a compression ratio of 8.0 to 1.

This new induction system with its better 'breathing' characteristics, results in increased acceleration and higher top speed. The fast acceleration is a safety factor which will permit extra-fast pickup when required to avoid traffic emergencies.

The 4-barrel carburetor is designed to provide a fast and free flow of air into the engine and a more thorough mixing of air-fuel mixture. It has two idle adjustments, one regulating the normal idle speed and the other the fast idle when the automatic choke is in operation.

In the lower speed ranges, only two barrels operate, but as additional acceleration is needed, the two extra barrels are brought into operation automatically. A choke interlock has been developed to prevent complete closing of the secondary barrels while the engine is being warmed up.

Exterior changes in the new models further emphasize the modern lines. A new gold 'V' ornament has been created for the grille and the same theme is used for side ornamentation and Lincoln is block lettered on the front.

A newly styled deck ornament and lock for the trunk compartment gives the car a wider and lower rear end appearance. A roof panel ornament between the rear door and back window on the 4-door models lends a hard-top look to these styles.

All Lincoln models have a one-piece rear window, adding to the wide appearance and providing greater rear view visibility.

Commemorating the 50th anniversary of the Ford Motor Company, a gold medallion has been mounted on the instrument panel.

Doors on the new Cosmopolitan and Capri have a two position check mechanism, including an intermediate stop that holds the door partly open when necessary in confined areas. This is an added safety and convenience factor which lessens the chances of doors closing on hands or legs.

Thirteen basic colors and 30 two-tone combinations with a variety of upholstery fabrics are offered.

Once again, Lincoln is available in four models: the Cosmopolitan sport coupe, the Capri 4-door sedan, the Capri hardtop and the Capri convertible.

In addition to improved models for '53, the Lincoln-Mercury plant has moved to their new factory at Wayne, Mich. to facilitate production.

The plant is the newest and most modern auto assembly unit of the Ford Motor Company. It is the fourth assembly plant to be built by the Lincoln-Mercury Division since 1948, when the Los Angeles, St. Louis and Metuchen, N.J. facilities were completed. ☆ ☆

An extra large trunk capacity is one of the outstanding features of the new models. Illustrating this are three men in space commonly used for luggage.

Dashboard and instrument appointments approximate last year's models. The wide-sweep speedometer is easy to read as are the other instruments.

The new model, with its 205 HP engine and quick acceleration easily outstrips last year's Lincoln in a special drag race.

1953 LINCOLN COSMOPOLITAN and CAPRI SPECIFICATIONS

Exterior:

Wheelbase	123 inches
Overall length	214.1 inches
Overall height, loaded	62.7 inches
Overall width	77.5 inches
Tread (front)	58.5 inches
Tread (rear)	58.5 inches
Curb weight (custom 4-door)	4,262 pounds
Tire size	8.00 x 15 super balloon
	8.20 x 15 on convertible

Interior:

	4-Door	Coupe	Convertible
Head room—front	35.5	34.1	35.1 inches
Head room—rear	34.7	33.4	35.3 inches
Leg room—front	44.3	44.3	44.3 inches
Leg room—rear	42.8	38.2	38.1 inches
Shoulder room—front	57.5	57.5	57.5 inches
Shoulder room—rear	57.2	55.5	46.0 inches
Hip room—front	62.3	62.3	62.3 inches

Engine:

Horsepower	205
Bore	3.80 inches
Stroke	3.50 inches
Displacement	317.5 cubic inches
Torque	305 pound-feet @ 2000 RPM
Compression Ratio	8.0 to 1

General:

Rear axle ratio	3.31
Rear axle (make)	Spicer
Drive	Hotchkiss
Brakes (diameter)	11 inches
Brake lining area	202.34 square inches
Turning diameter	45.3 feet
Front suspension	Ball-joint, individual coil
Rear springs	Longitudinal, semi-elliptical
Battery	63 plates, 110 ampere-hours
Water capacity	22.5 quarts
Fuel tank capacity	20 gallons

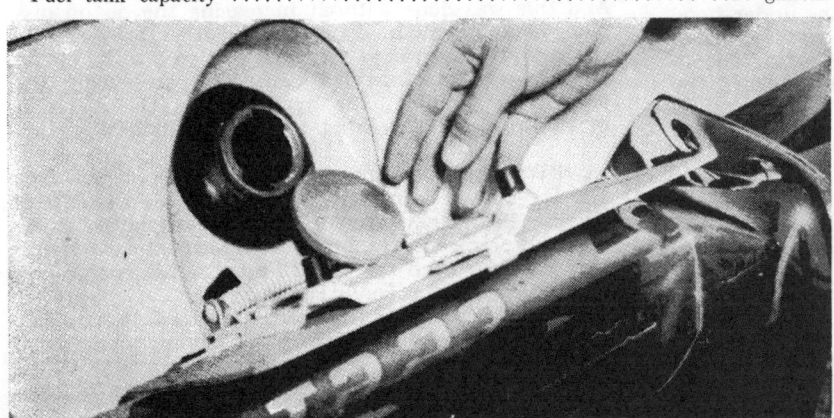

Above: Jack Bennett, engineering sales representative, adjusts the seat control mechanism. The gas filler cap (left) is behind the center-mounted license plate. Plate is attached to hinge unit.

LINCOLN FEATURES SIMPLICITY OF DESIGN, WITH ACCENT ON RUGGED FRONT END.

ROAD REPORT ON THE 1953 LINCOLN

By GEORGE BALASSES

THE 1953 Lincoln is a sleeper—and I don't mean the overnight kind. From the outside it looks much the same as the 1952 model, but the resemblance ends there.

Lincoln engineers have been busy in the under-hood department. I found this out as soon as I watched an acceleration race between a '52 and a '53 Lincoln at the Dearborn proving grounds. Later I drove the new model up hills, over bumpy roads, and around the speed track to see for myself what an easy-handling powerhouse the new Lincoln is.

In the aforementioned acceleration test I saw the '53 model leap ahead of its year-old brother with unfraternal zest in the first few feet and gain a lead of 200 feet in the course of a 1,000-foot run. Official acceleration figures backed up what I saw:
1952 Lincoln—0-60 m.p.h.—16 sec. 0-80 m.p.h.—30 sec.
1953 Lincoln—0-60 m.p.h.—12 sec. 0-80 m.p.h.—22 sec.

The top speed of the '53 Lincoln is 112 m.p.h. against 100 for the '52; and this seemed to be honest measuring, rather than optimistic waving by the speedometer needle.

Most startling fact about the engine is the increased horsepower—from a respectable 160 to a loaded 205. And this was done without increasing the piston displacement. Here's how Lincoln did it: (1) by raising the compression ratio from 7.5 to 1 to 8 to 1; (2) by installing a four-barreled carburetor (replacing the two-barrel '52 job); (3) by increasing the diameter of the intake valves from 1.74 to 1.98 inches.

ASIDE FROM the speed and power, several new features have been introduced to increase the ease of driving. Brand-new are the power brakes, power steering, and power seat raising—these make the car handling easy for even the tiniest woman, no matter how snarled the traffic conditions may be.

With power brakes, the pedal push effort has been reduced by one-third. When I was zipping along at 90, I touched the brake pedal and was surprised at the easy, even action which slowed the Lincoln without any slurring or grabbing. I found that I could keep my heel on the floor and easily swivel my foot from the accelerator to the brake. This is a considerable safety increase when a single second counts for 88 feet of travel at 60 m.p.h. Official figures show a 30 per cent saving in the amount of time needed to apply the brakes.

Power steering should be the little woman's particular delight, and should also be welcomed by strong men. Parking effort is reduced from a 35-to-40 pound hand pull on the steering wheel rim to a gentle 3-to-7-pound caress. Rounding a fairly sharp curve at 90, all I had to do was stroke the wheel and we eased around without a tremor. I felt not the slightest sign of wheel resistance in spite of the racing speed.

Power seat adjustment is a two-dimensional joy. Cold statistics reveal that there is a maximum 4½ inch horizontal travel and a vertical lift of 1½ inches. But no figures can transmit the luxury of flicking a switch and quickly adjusting the seat to the driver's exact preference, both forward and backward as well as up and down.

I NOTICED that Lincoln, along with a number of other manufacturers, has continued all-out for visibility. More vision should mean more safety. The back wraparound window affords a clean sweep of rear vision. The one-piece windshield is so wide and deep that I got the impression of sitting behind a "picture window."

Another feature I liked was the spread-out instrument panel. I had become accustomed to the bunched-up instrument arrangement and had forgotten that another grouping might be better. I could even see what time it was by the neat dash clock.

My only criticism concerns the body styling. By means of a little projection I can see where many people might not instantly recognize a Lincoln going past. Furthermore, there is even a possibility that the high class Lincoln might be mistaken for a Mercury or a Ford.

As the top-quality Ford product the Lincoln should stand out like a shining knight. Children should be able to take a glance and say, "There goes a Lincoln."

Aside from this one point the Lincoln seemed to me to be a luxury car that can hold its own in fast get-aways with smaller cars, and still provide the extra comfort, size and impressiveness of the most expensive models.

CONTINUED ON PAGE 91

Huge air filter almost covers powerful 205 h.p. engine. Power steering motor is located upper right.

Instrument panel features all-in-a-line design for easier reading, especially of speedometer.

43

ROAD and TRACK ROAD TEST No. A-1-53
1953 Lincoln Sedan

SPECIFICATIONS

Wheelbase	123.0 in.	Horsepower at 4200 rpm	205
Tread—front	58.5 in.	Compression ratio	8.0:1
—rear	58.5 in.	Gear ratios—4th	3.31
Tire size	8.00 x 15	—3rd	4.80
Curb weight as tested	4440 lbs.	—2nd	8.70
—front	2590 lbs.	—1st	12.64
—rear	1850 lbs.	Transmission	Hydramatic
Engine	V-8	Mph per 1000 rpm	25.5
Valve system	pushrod ohv	Mph at 2500 ft./min.	
Bore & stroke	3.80 x 3.50 in.	piston speed	110
Displacement	317.5 cu. in.	List price	$4790.12
Torque at 2650 rpm	308 ft. lbs.	(without extras)	$4015.13

TAPLEY READINGS

Pulling Power	Gear	Mph
600 lbs per ton	1st	19
500 lbs per ton	2nd	30
325 lbs per ton	3rd	46
203 lbs per ton	4th	72

Wind & Rolling Resistance (coasting)

40 lbs per ton at	10 mph
51 lbs per ton at	30 mph
94 lbs per ton at	60 mph

SHIFT POINTS
(Mph)

From	1st	2nd	3rd
Start in "4"	19	35	60
Start in "3"	19	35	70
Start in "LO"	19	38	70

(Manual shift to "3" at 38 mph)

PERFORMANCE

Test Conditions: Calm, cool night and morning, 60/70° F. Driver, passengers, equipment weight, 600 lbs.
Top speed (avg.) 108.05 mph
Fastest one way 109.80 mph

ACCELERATION

0-30 mph	6.0 secs.
0-40 mph	7.8 secs.
0-50 mph	10.7 secs.
0-60 mph	14.4 secs.
0-70 mph	19.3 secs.
0-80 mph	24.1 secs.
Standing ¼ mile	19.8 secs.
Fastest one way	19.4 secs.

SPEEDOMETER ERROR

Speedometer	Actual
10 mph	9.5
20 mph	18.6
30 mph	28.4
40 mph	37.5
50 mph	45.9
60 mph	53.6
70 mph	63.0
80 mph	72.8
90 mph	85.7
100 mph	95.7
110 mph	98.9

FUEL CONSUMPTION

13.5 to 17 mpg   (See text)

"America's Fastest Stock Car!"

In view of the overwhelming Lincoln victory in the Third Mexican Road Race (first 4 cars in the stock category) the prospect of a full scale road test of this 205 horsepower car promised some very interesting "work".

A week after the Mexican race, 1953 Lincoln's were hard to find in Southern California. However, through the cooperation of the Lincoln Factory Branch and Inglewood dealer, Bob Estes, we were loaned a "fully powered" Cosmopolitan sedan. "Fully powered" may require some explanation, it being Lincoln's description covering; Hydramatic transmission, power steering, power brakes, power seat, and power operated windows.

Great things were expected from this car, and we were not disappointed. Our tests were not run under the most favorable conditions; on the contrary, three points should be brought out. 1. The car had only 460 miles on the odometer when we took it over. 2. The Lincoln is geared for the best possible top speed and economy. This reduces the acceleration ability that might have been incorporated at some sacrifice in speed and economy. 3. The fully powered Cosmopolitan sedan weighs almost 250 lbs. more than a "bare" model, and as tested, our equipment and personnel added another 600 lbs. for a staggering total of 5040 lbs.

Top Speed . . .

During the speedometer calibration runs we found that after 90 mph. it took the Lincoln quite a while to build up speed. Our desert road is absolutely level and straight for over 3 miles on each side of the surveyed ¼ mile, and we needed all of this to "peg" the speedometer. With only 600 miles on the odometer (by then) we made two timed runs and then decided to give the new engine a cooling-off period, before making the third and fourth runs. However, at no time was there the slightest sign of overheating, a feature of the Lincoln which all of the Mexican race drivers praised. Unquestionably the 1953 Lincoln is the fastest stock car in America, and we are confident that with say 10,000 miles on the engine, the honest timed top speed of this car can be set at 112 to 115 mph.

Acceleration ...

The acceleration performance of the Lincoln is closely allied to the advantages and disadvantages of the Hydramatic transmission (standard equipment—no alternative). The gear ratios of the transmission are designed for broad contingencies and are too wide apart for best acceleration on level roads. As a result, considerable experimentation was necessary to evolve the best technique for the acceleration tests. Our method was as follows: rev the engine up to (as a guess) 1500 rpm, move the gear selector suddenly from N to LO, and as the car moves off, press the accelerator gradually to wide open. This gives the desired leap at the start, without too much wheelspin. At an indicated 40 mph move the lever from LO to 3, and leave it there. This position moves the 3 to 4 shift point up from a normal 66 mph (at W.O.T.* in 4) to an indicated 77 mph. With practice we gradually brought the time for the standing ¼ mile down to 19.4 secs.—this with a gross weight of over 5000 lbs. and on an early morning dew-damp surface.

Our acceleration figures to 60 and 80 mph (after speedometer correction) are just 2 seconds slower than factory claims, and we are sure that a car with more mileage, and carrying 600 lbs. less weight can live up to a zero to 60 mph time of 12 seconds, and a zero to 80 time of 22 seconds. These speeds, incidently, were in our tests, run to

———
* Wide-open throttle

1953 Lincoln Acceleration (thru the gears)

ROAD and TRACK

Hop Up Magazine's road test of a Mexican Race entrant was run concurrently with Road and Track's sedan. The race car was one second better on the standing quarter mile, and its best top speed run was 116.9 mph.

speedometer readings of 66 and 87 mph respectively.

Economy ....

The speed and performance of the 1953 Lincoln was expected, but the fuel consumption was a genuine surprise. Cruising at 60 to 70 mph we averaged 17.0 miles per gallon for 96 miles. Driving back from the desert test strip, and enjoying the car to the fullest, we averaged 13.5 mpg for 138 miles, even though cruising at an indicated 80 mph and hitting 90 mph occasionally.

Handling Qualities ...

As mentioned earlier, this car had both power brakes and power steering. The power brakes were the best we've yet tested, primarily because you can barely tell (if at all) that the booster is operating. Yet if you step immediately into a normal car the difference is amazing. Lincoln's secret, if any, is that pedal pressures and travel are reduced only to one-half that of non-boosted brakes, so eliminating the extreme sensitivity of some power-braking systems.

Unlike the brakes, our staff cannot praise the Lincoln's power steering. Frankly, no one liked it. While many women will find this steering advantageous for city driving, it leaves much to be desired for open highway travel. Not that it isn't safe—it is, even though it takes some practice to avoid trouble when a sharp turn is encountered unexpectedly. What we didn't like was that it proved very tiring to drive even 100 miles at a stretch on an average straight highway.

CONTINUED ON PAGE **49**

Capri '53 LINCOLN

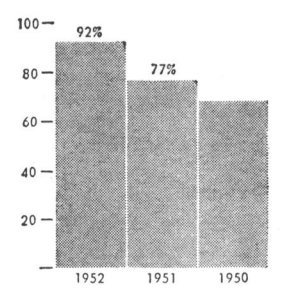

An MT Research Report

By Walt Woron
and
Pete Molson

DEPRECIATION

92% 1952
77% 1951
1950

FUEL CONSUMPTION

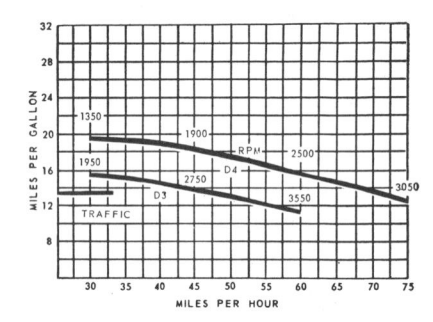

ACCELERATION

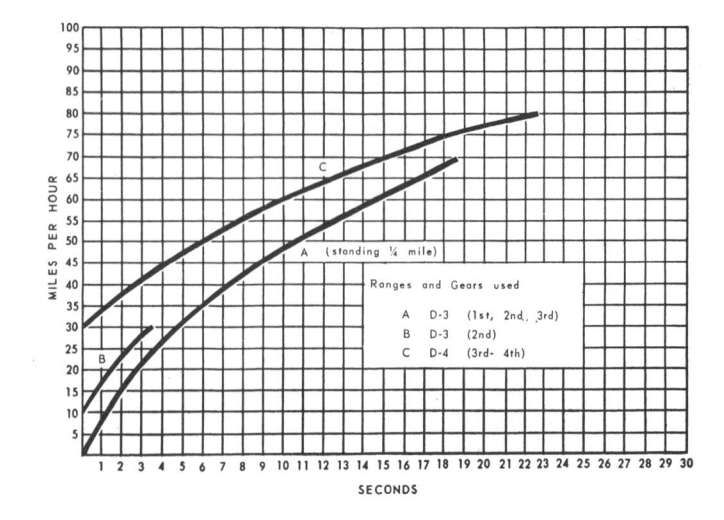

Ranges and Gears used
A D-3 (1st, 2nd, 3rd)
B D-3 (2nd)
C D-4 (3rd- 4th)

A (standing ¼ mile)

BOX SCORE

	Below Avg.	Avg.*	Above Avg.
Acceleration			
Standing ¼ mile			
30-60 mph			
Body work			
Brakes—stopping distances Avg. @ 30, 45, 60 mph			
Ease of handling			
Fuel economy—averages @ 30, 45, 60, and traffic			
Interior			
Ride			
Roadability			
Top speed—average			

*This box score is based on the average of all '52 cars tested except for the ratings on Interior and Ride, which refer to the average of other cars in the same price class (see February '53 MT).

Front compartment boasts plenty of room, high-grade fabrics, a well laid out dash that occasionally can dazzle the driver

46

Combines Luxury With Power

Photos by Jack Campbell

"THE TOUCH of casual magnificence," announces the Lincoln brochure ". . . Designed for the world you move in." If all of us do not move in worlds of casual magnificence, we must at least admit that the 1953 Lincoln is a modern and worth-while addition to the U.S. fine-car field.

At the time of its introduction the most powerful American automobile (205 hp), the Lincoln teams its capable engine with an air of compact but unstinting luxury and an as-yet-unsurpassed number of power-driven optional extras to give extreme driving ease. These include power steering (in the Lincoln, this is a cross between the Chrysler and GM types, using a fairly quick ratio but taking over only after four to seven pounds of effort are applied to the wheel); power braking, made doubly useful because of the low, pendulum-type brake pedal; electrically operated windows, and a four-way, button-operated front seat adjustment.

You soon become accustomed to such fine items but you have to pay for them. The sum total of the above accessories runs $419.30. If you can afford it, they are conveniences to be desired. The value that is placed on them will have to be your own.

What Lincoln has done is to take the all-new car introduced last year and refine it, upping the (advertised) hp by 45 and carrying out myriad, and generally well-considered, changes that should make the '53 model a much stronger contender for Cadillac's coveted spot as America's No. 1 prestige car. At first glance, the new Lincoln is hard to distinguish from the '52 model; it uses the same body and rugged grille-bumper combination, but has sprouted gold Vs here and there in honor of the Ford Motor Company's 50th anniversary. It is a handsome car, expensive looking, not at all bulky.

The large and heavy doors have push-button latches and stationary handles, and particularly lend themselves to the new two-position door check that appears on all Ford products this year. It is reasonably easy to slip in or out with the doors locked in the half-open spot, a particular blessing in tight parking lots and on busy streets.

The seat is unusually comfortable. The standard adjustment, mechanically operated, tilts the seat back forward as the seat moves toward the dash. One optional electric mechanism (including electric windows, which is the only way it can be had, it costs $139.80) does the same thing; the other, which is $177.40 with electric windows but can be bought by itself for $69.90, uses two buttons, one for height and one for distance from the controls. On the test car, which had the power-driven seat, Mrs. Woron glided forward by remote control. When it came my turn to drive, the power unit refused to operate, and I sat crowded against the wheel until the motor functioned again an hour later. This probably would not occur as a rule, but the fact that it *can* happen is against a strictly automatic device, particularly one which, like the Lincoln seat, cannot be operated manually.

Electric windows, which are available only in combination with one or the other of the automatic seats, can all be operated from the driver's door. They are quiet and their motors shut off automatically with the aid of an overload circuit breaker, signaling the driver when they reach the end of their travel. Electric windows are fine until something happens to a motor in the rain or in a snowstorm. Of course, this may never happen, but the chances of a breakdown are somewhat greater than with a crank. The standard cranks require 3½ turns to close or open a window completely. Quarter windows have cranks plus locks in the front compartment; in the rear they have the usual turn-and-push handles.

Lincoln designers have grouped a rather large number of controls on a handsome panel extending two-thirds of the way across the front compartment. Instruments are at the top, buttons and levers at the bottom and left; all are labeled, and the labels are illuminated at night. Large block style lettering and the positions of the instruments—two to the left of the speedometer and two to the right —make a sweeping glance suffice for checking. Heater and ventilating controls are, happily, an integral part of the panel; the ones on our car did not operate too easily and the defroster blower was quite noisy.

The glove compartment and ashtray both require a stretch from the driver. The glove compartment, of the door type, has a recessed push button, a good safety item. A large ridge around the instruments compares to many cars, further indicating a trend in panel design.

Dale Runyan, one of MT Research's experts on interiors, gives the consensus: "A

Editor Walt Woron examines tall, wide plastic taillights, biggest in the industry and a real safety feature at night

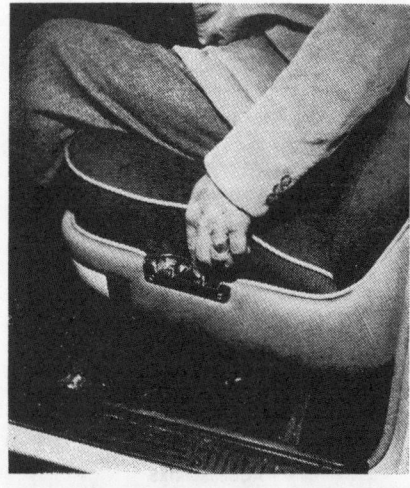

Perhaps the fanciest item among the power-driven equipment available for the Lincoln is the four-way, pushbutton seat

Lincoln's big luggage compartment has good-quality fabric lining to prevent scuffing. Gas filler cap is behind license

47

very nice trim job. Lincoln did not try to economize on materials or labor. Detailing is the equal of any American production car."

The switch and starter button are separate. A flasher operates when the ignition and the left-hand, cane-type hand brake are on. The Dual-Range Hydra-Matic dial—the automatic transmission is standard—is neatly faired into the steering column. The large overhead-valve V-8 engine, a bit reluctant to start and noisy at first, soon quiets down to a whisper. Even to our critical standards, it provides almost completely satisfactory acceleration at low speeds, at high speeds, or uphill. It has all the average driver can use, whether he wants to be first away from the signal or needs an extra margin for safe passing at cruising speeds. Acceleration does not appear to taper off, as with many cars; it just keeps coming.

Vision in the Lincoln is excellent. The huge one-piece curved windshield meets corner posts that are well placed and not too thick, and the huge one-piece curved rear window leaves no blind spots. The hood slopes downward at a good angle and the right front fender can be seen if one rises slightly. Unfortunately, the view forward and to the right is obstructed by the mirror, which in a right turn is situated squarely in the center of the driver's vision. In a luxury car, constant-speed electric wipers would seem more in keeping than the vacuum type.

Some glare is noticeable from the right side of the instrument panel, and some from the chrome trim on the steering wheel. The sun visors have small locks to keep them from rattling.

Thanks to its power steering, one can turn the Lincoln's wheels with very little effort, whether moving or stock-still. As the hydraulic mechanism takes over only after one has exerted four to seven pounds of pull, the tendency is to get set for a sudden lack of resistance. This sounds much more disconcerting than it is. After a two-day, 800-mile trip, it was no more apparent that the car had power steering, except for the lessening of fatigue. Lock to lock takes 3¾ turns, so that not only little effort but little actual movement is required to maneuver this 4400 pound car into a parking place. The right rear fender, as in most four-door sedans, cannot be seen from the driver's seat. Otherwise it is difficult to imagine how ease of parking could be much improved.

The brake and throttle are in near-perfect relationship. The wide brake pedal can be operated with the left foot if one prefers. With the usual method the operation is even simpler. All that is necessary is to pivot on the heel; the brakes can be applied with the ball of the foot. This

With the oversize air cleaner removed, MT's expert Ray Brown gets a chance to tinker with the four-barrel carburetor

is particularly easy with the power brakes, which use a pedal positioned closer to the floorboard.

The car is very docile and easy to handle. After the initial few pounds of effort, it goes around corners with a minimum of correction. Ordinary turns are fairly silent, but the large (8.00 x 15) tires give out loud screams at any sudden movement of the wheel, even at low speeds. More pressure in the tires would minimize this and also give the tires longer life. The big tires are also partly responsible for whipping aside on streetcar tracks.

1953 LINCOLN ROAD TEST TABLE

PERFORMANCE

CLAYTON CHASSIS DYNAMOMETER TEST

(All tests made under full load conditions)

RPM	MPH	ROAD HP
1200	31	46
1500	37	56
2000	54	82
3100	83	112

Per cent of advertised hp delivered to driving wheels—54.6

TOP SPEED (MPH)

(Clocked speeds over surveyed ¼ mile)

Fastest one-way run	112.70
Slowest one-way run	107.78
Average of four runs	110.42

BRAKE STOPPING DISTANCE

(Checked with electrically actuated detonator)

Stopping distance at:

30 mph	53 ft. 3 in.
45 mph	119 ft. 8 in.
60 mph	234 ft. 0 in.

SPEEDOMETER ERROR

Actual MPH	Indicated MPH
30	34
45	52
60	70

GENERAL SPECIFICATIONS

ENGINE

Type	Overhead valve V-8
Bore and stroke	3.8 x 3.5 in.
Stroke/bore ratio	.92:1
Compression ratio	8.0:1
Displacement	317.5 cu. in.
Advertised bhp	205 at 4200 rpm
Piston travel @ max. bhp	2450 ft. per min.
Bhp per cu. in.	0.64
Maximum torque	308 lbs.-ft. @ 2650 rpm
Maximum bmep	146.28 psi

DRIVE SYSTEM

Transmission: Dual-Range Hydra-Matic
Ratios: Reverse—4.30
1st—3.81 3rd—1.45
2nd—2.63 4th—1.00

Rear axle: Semi-floating Hotchkiss drive, hypoid gears. Ratio: 3.31 to 1

DIMENSIONS

Wheelbase	123 in.
Tread	front & rear 58½ in.
Wheelbase/tread ratio	2.1:1
Overall width	76¼ in.
Overall length	214⅛ in.
Overall height	62⅝ in.
Turning diameter	45 ft. 4 in.
Turns lock to lock	3¾
Weight (test car)	4440 lbs.
Weight /bhp ratio	21.6:1
Weight/road hp ratio	39.6:1
Weight distribution	front 57.7%, rear 42.3%
Weight/sq. in. of brake lining area	21.9 lbs.

INTERIOR SAFETY CHECK CHART

QUESTION	YES	NO
1. Blind spot at left windshield post at a minimum?	X	
2. Vision to right rear satisfactory?	X	
3. Positive lock to prevent doors from being opened from inside?	X	
4. Does adjustable front seat lock securely in place?	X	
5. Minimum of projections on dashboard face?	X	
6. Are cigarette lighter and ash tray both located conveniently for driver?		X
7. Is rear vision mirror positioned so as not to cause blind spot for driver?		X
8. Is the windshield free from objectionable reflections at night?	X	
9. Is the dash free from annoying reflections?	X	
10. Are the tires required to support only a safe amount of weight with normal loads?		X
TOTAL FOR LINCOLN CAPRI	70%	

PRICES (Delivered in Detroit)

Cosmopolitan:	Sport Coupe	$3762.43
	Four Door Sedan	3654.63
Capri:	Sport Coupe	4017.78
	Four Door Sedan	3909.98
	Convertible	4186.78

ACCESSORIES

Radio with vacuum antenna & rear speaker	$131.70
Dual heating system with defrosters	121.00
Road lamps	37.10
Windshield washer	11.27
Curb buffers (standard on Capri)	26.02
All leather trim on Capri closed models	96.80
Tinted glass	29.60
White wall tires (5)	36.10
Power steering	198.90
Power brakes	43.00
Electric windows & 2-way seat adjustment	139.80
Electric windows & 4-way seat adjustment	177.40
Electric 4-way seat adjustment alone	69.90

OPERATING COST PER MILE ANALYSIS

(In this portion of the test table, MOTOR TREND includes those items that can be figured with reasonable accuracy on a comparative basis. The costs given here are not intended as an absolute quide to the cost of operating a particular make of car, or a particular car within that make. Depreciation is not included.)

1. Cost of gasoline—premium	$157.78
2. Cost of insurance	168.40
3. Maintenance:	
a. Wheel alignment	4.55
b. ½ brake reline	18.88
c. Major tune-up (one)	10.50*
d. Automatic transmission, adjust, change lubricant	17.85
First year cost of operation per mile	3.8c

MAINTENANCE AND REPAIR COST ANALYSIS

(These are prices for parts and labor required in various repairs and replacements. Your car may require all of them in a short time, or it may require none. However, a comparison of prices for these sample operations in various makes is often enlightening.)

Part	Cost	Labor
1. Distributor	$11.95	$ 3.50
2. Battery	29.95	1.25
3. Fuel pump	17.48	1.40
4. Fan belt	2.16	1.05
5. Valve grind	2.00	33.60
6. One front fender	57.05	23.45
7. Two tires	61.72*	
8. One bumper	61.90	5.60
TOTALS	$244.21	$69.85

*Note: MOTOR TREND constantly revises its test procedures and methods of analyzing probable costs of ownership in an effort to arrive at the most realistic figure. This month, the cost of one major tune-up is listed instead of two, and the cost of two tires replaces the cost of one.

Lincoln Road Test

Steering wheel shock has almost disappeared from the Lincoln, even on washboard roads. One can feel some vibration on rough surfaces, but no solid shock. The front end never feels mushy, and the car, with only 42 per cent of its weight on the rear wheels, breaks loose on corners only in soft dirt or gravel. However, in this case one goes into a safe, controlled slide. At higher speeds, it is necessary to correct for both wind wander and sudden cross gusts, a characteristic of practically all cars; however, the Lincoln, because of its longer wheelbase and heavier weight, actually requires less correction on this score than many cars.

The medium-soft ride has a floating quality on average roads. Certain road and speed combinations produce an annoying steady bounce. The shock absorbers are in control at all times, bringing quick recovery after dips or bumps. The car seems to absorb shock, smoothing out the road considerably on washboard surfaces. Good suspension and the large tires absorb most of the shock usually felt over chuck holes. Aside from the persistent squealing on corners, the tires make themselves heard only over grooved or brick surfaces. "Walking," the tendency for the wheels to touch the road intermittently, is not excessive on ordinary paved surfaces. Body sway is not sufficient to cause passengers discomfort at any time.

Under the Hood

The Lincoln's horsepower has been upped from 160 to 205 as a result of several changes. There are a new four-barrel carburetor, a new camshaft and new intake valves with greater clearance. Redesigned combustion chambers have raised the ratio to 8 to 1. New exhaust manifolding has reduced back pressure, and higher axle ratios give more horsepower at the rear wheels. All this resulted in the highest average top speed that MT Research has yet recorded (110.42 mph) and in an enviable record in the Mexican Road Race (see MOTOR TREND for January and February).

Ray Brown, MT's engine expert, joined the research team for investigation inside the engine compartment. The manifolds, the big air cleaner and the right-side ventilating duct and blower effectively conceal much of the engine. The spark plugs are completely out of sight, but removal of the ventilating equipment, mounted with quick-release catches, allows access to the right-hand bank and to the starter solenoid. Even so, plugs must be removed with the aid of an extension and by the touch system.

As in all new Ford products, the brake master cylinder is mounted on the firewall, and the Lincoln's brake booster is there, too. Removal of the air filter makes it possible, if not easy, to reach the distributor. Accessibility to the points and condenser is fair, but removal of the distributor assembly is cumbersome.

Get Out and Get Under

Last year, the Lincoln's battery returned to a spot under the floor, and there it remains, shielded, as the factory points out, from engine heat. It is somewhat handier for servicemen under the floor, but more easily forgotten there, and since Lincoln still uses a six-volt system even with all its electrical attachments, the battery should be where it will receive the constant attention any heavily loaded battery needs.

Despite this and the position of some of the other components, the Lincoln does not rate low on the accessibility scale. For example, it is possible, as on other new Ford engines, to check and add oil without changing clothes.

Potential Lincoln owners are unlikely to include many back-yard mechanics, but having to put the car on a hoist to remove the starter and generator is not conducive to low service charges.

Ray Brown adds, "This one is a natural for a dual-point distributor plate. I cannot understand why the factory does not take advantage of this simple method to insure longer life for the contacts by increasing the cam angle."

Mobilgas fuel (premium in this case, because of the high compression ratio) was used, as in all MT Research road tests. Acceleration (see graph) closely approximated that of the '52 Cadillac and, of course, was faster than the '52 Lincoln. High-speed acceleration was especially satisfying.

Fuel consumption averaged 1.3 mpg less than the average for all '52 cars tested and only a fraction of a gallon less than that for the '52 Lincoln. Considering the car's weight, this shows a high potential for one of the top places in this year's Mobilgas Economy Run on the basis of ton miles per gallon. Stopping distances were somewhat greater than average.

After approximately 2000 test miles, the Lincoln showed no appreciable tendency to loosen up. It was free from rattles and unexplained noises, the quietest of the '53 models we have yet driven. One valve tappet was noisy and the power booster for the brakes occasionally failed to operate. The warning light, intended to operate when ignition and parking brake are both on, was wired backwards on the test car, and flashed when the brake was off. These faults were felt to be the result of improper initial adjustment rather than of the shakedown.

The Lincoln comes close to the present-day top as a production car, intended for family use in the United States under various driving conditions. Men will welcome its eager power under all conditions. Women will appreciate the car's ease and comfort of handling, and its atmosphere of thoroughly enjoyable luxury in a reasonably sized package.

—*Walt Woron and Pete Molson*

CONTINUED FROM PAGE 45

There is a normal feel of the road, and one experiences no worry even at speeds of over 100 mph, but somehow it seemed to require a lot of effort to do the normal see-saw movements required of any car for straight driving. We also felt that 4 turns of the wheel for full left to right lock was one too many for any car, and especially so for a power steerer.

We also had the opportunity to try (for a few miles) an actual Mexican race entrant of the same make, equipped with heavy duty springs, shock-absorbers, and normal steering. Compared to our test car there is no denying that:

1. Top speed was better by 6 mph.
2. Acceleration was about equal.
3. The car rolled very little in a turn.
4. It rode like a truck.

The stock Lincoln feels almost too softly sprung at moderate speeds over rough roads, but it surprised everyone with the way it smooths out at 80 mph and above. The cruising speed on U.S. highways can be governed by driving conditions alone, except that at 90 mph the wind noise becomes noticeable and at over an honest 100 the noise is a shrill shriek.

General Comments . . .

The 205 bhp Lincoln ohv V8 engine is an outstanding engine, anyway you look at it. Smaller than its other two competitors (we needn't name them) it is exceptionally quiet and smooth at all times. Despite a compression ratio of 8.0 to 1 and an intake valve as big as the old F-head Hudson Super 6, it gives its power without any sign of detonation on pump ethyl gasoline. One item that might be over looked is the unusual combination of the highest torque peaking speed of any American car, (2650 rpm) coupled with an axle ratio of only 3.31. Such a pair would be absolutely incompatible without the dual range Hydra-matic transmission. Another interesting point is that the bhp peaking speed (4200 rpm) coincides with the top speed, indicating a perfect choice of gearing—at least insofar as obtaining best possible top speed is concerned. A third point is that cruising speed based on the usual factor of 2500 ft. per minute piston speed (as being safe for long periods) is not reached until 4300 rpm, or 110 mph. Thus the car can be cruised at any speed desired, *up to the maximum*, without concern for the engine.

For a car in this price category, the interiors are not impressive, though there are many niceties. The instrument panel layout is good, and the lighting system is only kept from being perfect by some reflection in the windshield when on "full bright". Dimming the dash lights eliminates the reflections and the instruments are still readable, once you learn which is which. The 4-way power adjusted seat is a never-ending source of amusement as well as utility, but the power operated windows would not be kind to out-stretched necks.

Visibility is excellent, particularly for the driver, who can see both fenders as well as the road only a few feet ahead of the car. The car is not actually very large overall, but it does take quite a little judging experience before you can confidently travel in dense traffic or judge parked-car clearances.

Summed up, the 1953 Lincoln, with its exceptional performance combined with surprising economy, seems destined to polish the reputation that somehow got slightly tarnished by the "light breeze". JB •

LINCOLN 205

Speed Age Staff Research Report

Capri

The performance of the 1953 Lincoln has set the country talking, just as did the design of the immortal Continental in 1938.

LINCOLN'S history, design-wise, is probably one of the most radical of any car on the market today.

From the time the Lincoln Company was formed in 1917 to build Liberty airplane motors for the Government, the company's line has ranged from one of the biggest engines in the business to the current much talked about model.

Actual production of Lincoln automobiles began late in 1920, as rugged and homely as the man it was named for. But, a little more than a year later, feeling the full effect of a depression, the plant was ready to close its doors. It was then Henry and Edsel Ford entered the picture. In January of 1922 the Lincoln Company was bought for $8,000,000.

Production of the Lincoln that year stood at 35 per day. Its big 358 cubic-inch V8 engine was 40.5 cubic inches larger than the 1953 Lincoln.

The 180-degree, high-torque crankshaft was machined all over and the compression ratio stood at 4.81 to 1 compared to 8.0 to 1 in the '53 model.

Bore, stroke and horsepower were unchanged from 1921 through 1927; then in 1928, the engineers got five more horsepower by increasing the bore an eighth of an inch and a vibration damper and crankshaft counterweights were added to smooth out the roughness.

A 5-passenger sedan that weighed a little over 5,000 pounds had a top gear acceleration from 10 to 30 MPH of 7⅘ seconds compared to today's sedan of 4,262 pounds with its acceleration from 0-30 in 3½ seconds.

Through the years the Lincoln has been known for its fast, steady cruising ability. The suspension was solid and fast corner-

Dick Adams Photo

This beautiful Capri hard-top is one of five models offered by Lincoln for 1953. Others include the Capri 4-door sedan, Capri convertible, Cosmopolitan sport coupe and Cosmopolitan 4-door sedan. Note fiftieth anniversary 'V' over bumper.

A profile of the Capri reveals a rather deadly spear on the hood.

Test driver Vince McDonald drives the Lincoln down the beach at Daytona to get a top speed reading. True top speed on one run was 109 MPH.

ing with very little sway or roll made this a luxury car of the day.

In 1931 a new cam and a dual-throat downdraft carburetor were added and the compression ratio was jumped from 4.81 to 5.25 to 1. This new model weighed about 5,500 pounds, enough to offset any great gain in acceleration, although top speed increased about 5 MPH.

The V8 engine designed by Henry M. Leland, the Lincoln Company's founder, disappeared in 1932 and was replaced by the first of the V12's. A 448-cubic inch job, known as the 'Big 12', this displacement giant was made for two years; the car it powered sold f.o.b. at the factory for anywhere from $4,300 to $7,200. On today's market with its inflated dollar, this would mean a price tag of $8,000 to $15,000.

This model only averaged eight miles to a gallon of gas, but could do a 0-60 in top gear in 26 seconds and a flying half mile at 95.75 MPH. At 30 MPH, the vacuum boosted brakes would stop this monster of almost three tons in 28 feet.

In 1933, the little twelve—a companion to the 'Big 12'—with a 382 cubic-inch displacement was born, but Lincoln was falling victim to a new depression. By 1935 production had dropped to 10 cars a day and the quality that had been built into the engines was giving way to less expensive methods. Performance, when new, was equal to its predecessors, but the longevity was gone.

About this time, the Lincoln Zephyr appeared. Originally designed as a rear engined car, it was produced as a conventional model when it became evident the cost would be prohibitive. The Zephyr, in 1935, was singled out as the first successful streamlined car in America. Powered by a 12-cylinder, 110 HP engine, its top speed was approximately 87 MPH.

The Continental was born in 1938 and was Edsel's brainchild. When it went into production the rich and famous adopted it almost immediately as the prestige car. It wasn't faster than its competitors in the top price bracket, but its cornering ability and distinction were unequaled. During the post-war period, the price of this model went so high that one used car dealer sold one for $10,000.

The Continental of 1948 was the last of that line, the company bowing to the high cost involved in producing an automobile requiring so many specially assembled body parts.

Last year, Lincoln served up a change of pace by abandoning the bulging body in favor of a sleek, slender sheath. Under the hood was an all new, overhead valve V8 engine rated at 160 HP and hailed by many as having a terrific potential.

That the potential was there is evident in the bomb Lincoln engineers set off when the 1953 prototypes' tests had been completed. With the exception of an unobstructed wrap-around rear window, a 2-stage door latch, and minor trim changes, this year's model resembles its predecessor. But under the hood, that 160 HP has been boosted to 205, the second highest in the industry.

The stepup in horsepower is attributed to a redesigned intake manifold, larger intake valves, reshaping of the combustion chamber, a straight-through exhaust system, a 4-barrel carburetor and improved air cleaner. The compression ratio has been increased from 7.5 to 8.0 to 1. This increased horsepower and resulting speed goes well with the handling qualities of the car.

The 1953 Lincoln tested by SPEED AGE was obtained from Warren A. Bushey, Administrator-Manager of Lincoln-Mercury Division, Washington District Sales Office, who graciously relinquished his Capri hard-top coupe for almost two weeks while the SPEED AGE staff drove it to Daytona Beach, Fla., and back.

The explosive, quiet power of this car was exhilarating over the relatively flat country between Washington and Daytona Beach.

The '53 Lincoln retains the same frame and suspension system that was brought out in the '52 car, a radical change from former years and something new in American production cars. Box channel side

Good visibility offered by the hard-top is apparent. The rear corner post interferes little with driver vision and blind spots are practically non-existent.

rails, and an X brace with five cross members makes this a very rigid frame.

The front suspension has two ball and socket joints on each wheel, one connects the top and the other to the lower support arm, eliminating the king pin support and allowing vertical motion to absorb road shock.

The automobile's low center of gravity and handling characteristics continue far past the point where a good many other models leave off.

Safetywise the Lincoln will compare with anything on the road today. Its big, powerful engine will very seldom, if ever, be used at top speed. But even the most careful driver occasionally finds himself in trouble where a quick surge of power is the necessary safety factor.

Driving over the pitching roads of Georgia in a blinding rainstorm at speeds as fast as any car traveling at that time, the power and ease of handling was unbelievable. Regardless of how hard the Lincoln was put through the turns, it failed to break loose and there was no discomforting lift and roll.

After driving the new Lincoln nearly 2,500 miles, it is not too hard to understand why this car swept the first four places in the stock division of the Pan American Road Race.

The test car was equipped with power steering and required only a four to seven pound pull to turn the wheel. This coupled with much faster steering (only 3¾ turns lock to lock) led to taking all the bends square the first 100 miles or so out of Washington. In city traffic power steering is a welcome addition to the extras offered by the automotive manufacturers. However, on the highway at speed, it takes a while for the driver to get accustomed to its quick, light positive action.

Power braking is another convenient addition. Mounted from a bracket under the dash, the foot pedal is approximately the same height from the floor as the accelerator and can be reached by merely pivoting the ball of the foot. With this unit, braking pressure is reduced two thirds.

The hand brake, which works independent of the regular braking system, is capable of locking both rear wheels when set. Its position is convenient to the operator's left hand, yet, when on, allows free passage in and of the door.

The interior is simple, but rich. However, although definitely a luxury automobile, the Lincoln does not list air conditioning as optional equipment. All instruments on the panel are easily visible and all controls within reach as is the glove compartment. Some panel illumination is reflected in the windshield and the intensity of the high beam indicator is annoying.

The ash trays on each arm rest in the rear are equipped with cigarette lighters; however, the trays on the test car were of little use. Every time the cover was opened, a terrific draft blew ashes out of the tray. This draft also comes through the back seat when the center arm rest is lowered.

The front seat position can be controlled by two fingertip buttons on the operator's side. Lincoln's power seat is driven by an electric motor and adjusts to a position to please even the most discriminating motorist. It will move as much as one and a half inches vertically and four inches horizontally.

It's a fact that altering the seat position every 75 to 100 miles makes a long trip less tiring. Now it's possible to do so without the driver's eyes leaving the road.

The engine and its accessories was never intended to be worked on by the amateur. The battery has been returned to a position under the floorboard. This, alone, could discourage attention and—unless the car is serviced by someone you can trust—the battery could easily become a forgotten item. Even the spark plugs, located under the manifolds, would be hazardous to change if the engine was hot.

The '53 Lincoln, as did the '52, has a low flat silhouette with both front fenders visible from the driver's position. This is a little frightening at first, for with most cars the position of the right side of the automobile on the highway is a matter of judgment and guess work. Being able to see the right front fender breeds a new sense of proximity and, at first, a tendency to shy away.

The low, long lines would be distinctive if it were not for the other cars in the

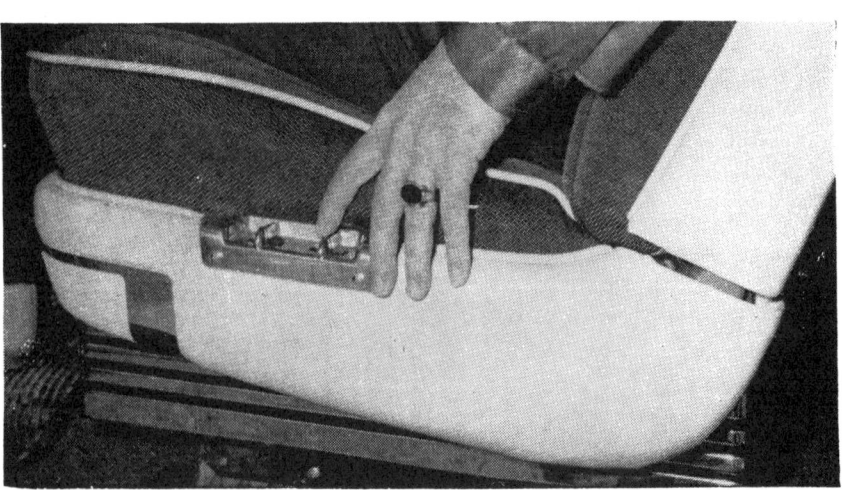

Don O'Reilly Photo

A big feature on the 1953 Lincoln for luxury-lovers is the electrically-operated front seat. A touch of the button will move the seat vertically and horizontally.

52

Ford family. Both the Ford and Mercury have been brought along to so closely resemble the Lincoln that, in some cases, when meeting these cars head on, it takes more than a quick glance to tell them apart.

Close examination of the body proved a little disappointing. An inspection of a dozen or so of the hardtop models disclosed, in every case, a separation just above the door on the operator's side and inside the rain gutter where the corner support fastened to the top. The continual opening and closing of the heavy door could very well be too much weight for the amount of bracing at this point.

Body workmanship was not quite what one might expect and there were evidences of re-doing in many places, where a rather poor job of retouching with paint was apparent.

Overall gas mileage, for nearly 2,500 miles was about 16½ miles per gallon. This is quite a contrast to the 10 miles per gallon attained by the early '30 models.

The performance of the test car could have been increased appreciably had the carburetor been adjusted. However, the Zenith milage tester showed some amazing figures. At 40 MPH, over a 2-way run, the average stood at 22 miles per gallon. But, in heavy city traffic the average dropped to 13½ miles per gallon. This compares favorably with much lighter and smaller cars in the field today.

At 80 MPH the average was 13½ and at 50 shot up to 17½ miles per gallon. This economy has been foreign to previous models.

Braking as tabulated on the Perf-O-Meter stood at 78%. This means that at 50 MPH, from the time the driver's foot was placed on the brake pedal, approximately 10 car lengths are needed to stop. Of course, depending on driver reaction time, up to an additional 10 car lengths could be traveled before the Lincoln was brought to a standstill. In emergency stops, there is no uncomfortable lifting or swerving.

From the gold V ornament on the front to the gasoline filler cap concealed behind the license plate in the rear, the Lincoln suggests power and ruggedness. With a performance that equals anything on the road, a roomy interior which will accept six people comfortably and a trunk that is almost large enough to pitch a tent in, Lincoln is going all out to regain first place in its price class.

Dick Adams Photo

The 205 HP Lincoln V8 which powered the first four finishers in the Mexican Road Race. Accessibility of virtually all parts of the engine is extremely limited.

1953 LINCOLN COSMOPOLITAN AND CAPRI SPECIFICATIONS

EXTERIOR:
Wheelbase	123 inches
Overall length	214.1 inches
Overall height, loaded	62.7 inches
Overall width	77.5 inches
Tread (front)	58.5 inches
(rear)	58.5 inches
Curb weight (custom 4-door)	4,262 pounds
Tire size	8.00 X 15 super balloon
	8.20 X 15 on convertible

INTERIOR:
	4-door	Coupe	Convertible
Headroom—front	35.5	34.1	35.1 inches
Headroom—rear	34.7	33.4	35.3 inches
Leg room—front	44.3	44.3	44.3 inches
Leg room—rear	42.8	38.2	38.1 inches
Shoulder room—front	57.5	57.5	57.5 inches
Shoulder room—rear	57.2	55.5	46.0 inches
Hip room—front	62.3	62.3	62.3 inches

ENGINE:
Horsepower	205
Bore	3.80 inches
Stroke	3.50 inches
Displacement	317.5 cubic inches
Torque	305 pound-feet @2,000 rpm
Compression Ratio	8.0 to 1

GENERAL:
Rear axle ratio	3.31
Rear axle (make)	Spicer
Drive	Hotchkiss
Brakes (diameter)	11 inches
Brake lining area	202.34 square inches
Turning diameter	45.3 feet
Lock to lock Power steering	3¾ turns
Conventional steering	5 turns
Front suspension	Ball-joint, individual coil
Rear springs	Longitudinal, semi-elliptical
Battery	63 plates, 110 ampere-hours
Water capacity	22.5 quarts
Fuel tank capacity	20 gallons

Performance Data
Rate of Acceleration
0-10 MPH	2½ seconds
0-30 MPH	3½ seconds
0-60 MPH	11 seconds
40-60 MPH—3rd gear	7 seconds
40-60 MPH—4th gear	7½ seconds

Top Speed Runs
	Indicated	Actual
North-South Run	120 MPH	109 MPH
Opposite Direction	117 MPH	106 MPH
Average		107.5 MPH

Brakes
Efficiency 78%

Mileage Tests
City Traffic	13½ miles per gallon
80 MPH	13½ miles per gallon
70 MPH	15 miles per gallon
60 MPH	15½ miles per gallon
50 MPH	17½ miles per gallon
40 MPH	22 miles per gallon

Suggested Retail Price List
Cosmopolitan Custom Series
Hardtop Coupe	$3,782.10
4-Door Sedan	$3,679.10

Capri Special Custom Series
Hardtop Coupe	$4,026.10
4-Door Sedan	$3,923.10
Convertible	$4,191.60

(Above price includes Hydramatic Transmission)

Optional Equipment and Accessories
Power Steering	$198.90
Power Brakes	$ 43.00
Power Seat, 4-way	$ 69.90
Electric Window Lifts and 2-way seat	$139.80
Electric Window Lifts and 4-way seat	$177.40
Radio and Rear Seat Speaker	$131.70
Heater and Defroster	$121.00
Curb buffers (except Capri Models)	$ 26.02
Windshield washers	$ 11.27
Leather upholstery (Capri Hard Top Coupe onlyl)	$ 96.80
Tinted glass	$ 29.60
Road lamps	$ 37.10
8.00x15 Whitewall tires	$36.10
8.20x15 Whitewall tires (convertible only)	$ 36.76

☆ ☆

A Refined Lincoln for 1954 . . .

CAPRI 4-door sedan shows trim lines of Lincoln styling. All models are powered by Lincoln's OHV 205-h.p. V-8 engine. Upholstery is in gabardine, nylon, whipcord or leather.

WHILE Lincoln models for 1954, introduced in dealers' showrooms December 3, retain the proven 205-horsepower overhead valve V-8 engine, a considerable number (the company terms it a "multitude") of mechanical improvements and styling refinements have been made in the new line.

Included among the many significant engine refinements are a larger, more flexible single-diaphragm vacuum distributor control, a new venting action for the new vacuum-controlled four-barrel carburetor, redesigned hydraulic tappets, a new filter element in the fuel pump and a self-cleaning filter for the fuel tank.

Models offered are the Lincoln four-door sedan and "hardtop" coupe, and the Lincoln Capri four-door sedan, "hardtop" and convertible.

A new styling theme has been introduced on front and sides. It includes a new bumper guard air scoop and addition of three vertical bars located between the upper and lower impact areas.

Further emphasizing the car's wider appearance is what the company calls "functionally correct" outboard position of the parking lights and turn indicators. A new hood ornament with a wide "V", is background for the traditional Lincoln crest.

Jutting bumper guards, full wraparound bumpers, restyled side moulding and newly-designed rear quarter gravel shield are further 1954 styling identifications.

Integral back-up lights have been combined with the readily identifiable Lincoln tail lights as standard equipment on all models.

New interior trim schemes utilize a line of modern interior fabrics, including gabardine which is being offered as an upholstery fabric for the first time, whipcords, genuine leathers, modern weaves and spun nylon. Interior trims are color-harmonized with the exterior colors.

Larger brakes have been developed for the 1954 Lincoln, with brake diameter increased to 12 inches. This has resulted in a 10% increase in the braking area to 220 square inches. This additional capacity is effective throughout the entire car braking range, requiring a quarter to a third less effort to stop the car at given speeds.

A noise suppression feature has been incorporated on all tires for the Lincoln to minimize tire squeal caused by the vibration of the outside shoulder ribs of the tires when the car is driven around a turn fast enough to cause these

MASSIVE wrap-around bumpers with jutting guards and bumper guard air-scoop characterize 1954 Lincolns. Hydra-Matic drive is standard on all models including the Capri convertible shown here.

ribs to slide. Buttons or spacers have been added between the ribs of the tires to muffle the noise.

Greater rigidity in the hood and the front end has been obtained by a new, straight hood lock support plate rod, the lower end of which is attached directly to the frame.

The steering column on the 1954 Lincoln has also been made more rigid by the addition of a rod extending from the column support to the cowl. In addition, the area of the column supports has been reinforced by adding a brace to the dash and cowl top panels.

Power steering, power brakes, electric window lifts and four-way power seats are offered as optional equipment. Automatic transmission is, of course, standard equipment.

The Lincoln V8 engine has a compression ratio of 8 to 1 and a displacement of 317 cubic inches.

The four-barrel carburetor has been improved by development of a vacuum method of operating the second two barrels. It is equipped with a new external venting action that lets fuel vapors pass harmlessly outside the carburetor when the engine is idling or stopped, leaving the intake manifold dry and ready for instant starting. When the throttle is opened the vent is shut, permitting normal carburetor operation.

Development of a new throttle linkage has resulted in a smoother shift of the automatic transmission. The greater sensitivity of the new vacuum distributor control diaphragm eliminates the need for the mechanical linkage previously used and assures ample initial spark advance for smooth, positive acceleration from standing starts.

A low-restriction oil bath air cleaner forms the carburetor air horn and encloses a dual-float concentric fuel bowl which is mounted over the four barrels.

A new magnetic fuel pump filter in the 1954 Lincoln has nearly twice the efficiency of the former one, giving positive protection against the penetration of even the smallest metallic particles into the fuel system.

Hydraulic tappets have been redesigned to provide a higher oil level so there is plenty of oil to fill the compression chamber when starting a cold engine. ★

Five days that confirmed a motor car revolution

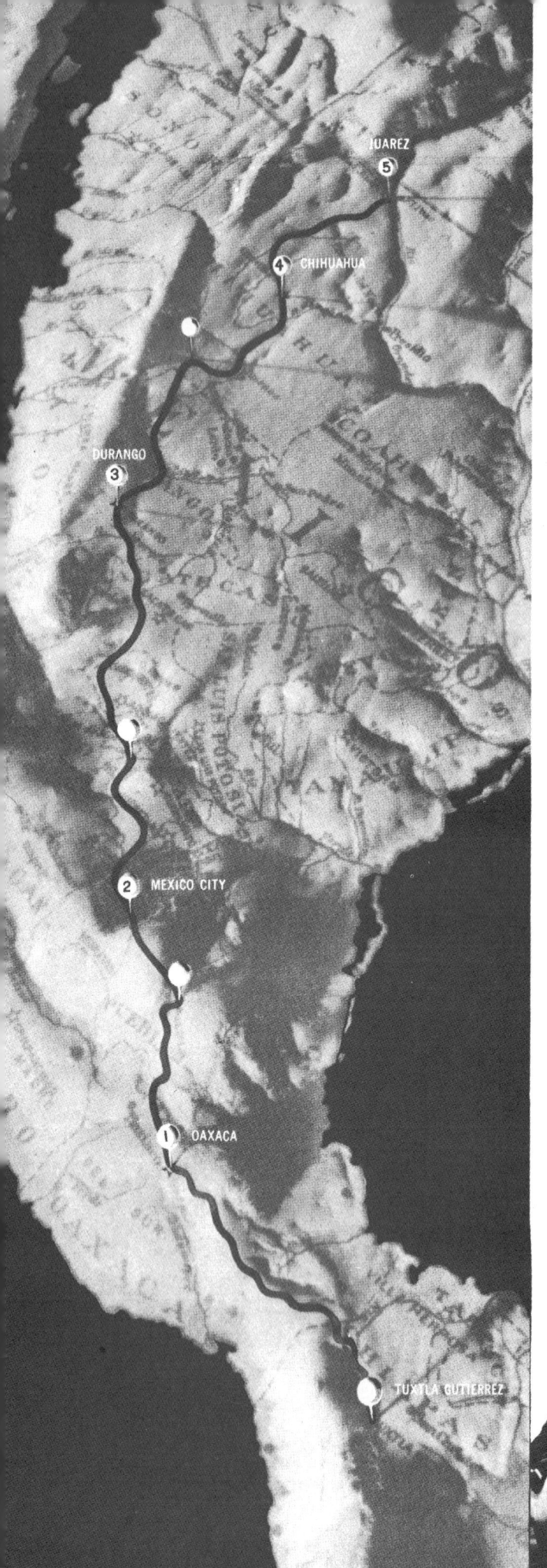

the challenge:

On Wednesday, November 19, 1952, expert drivers piloting the world's finest automobiles, lined up for the start of the world's toughest competitive race.

In two classes, stock and sport, for five days, they were to pit skill and machines against 1,938 miles of road from Tuxtla Gutierrez near Mexico's southern border almost to the international boundary at Juarez.

This was the route—and the field—an as yet publicly untested car, the 1953 Lincoln, set out to conquer. From near sea level to altitudes over 10,000 feet, it sought to best the world's finest stock cars and dared to contest the performance of the world's most costly, competition-inspired sports cars.

the result:

Even before the first of the four Lincolns crossed the finish line to take first, second, third and fourth places at Juarez five days later, there had been more than an inkling of the results to come.

They had consistently outperformed all other stock cars. They had rolled on and on—while 56 of the original 95 starters dropped out.

Besides sweeping the first four places in the stock car field, Lincolns outran or outlasted 76% of all sports cars—cars specially built for competitions like this!

Truly a motor car revolution had been confirmed. And here, on the following pages, as told by the drivers themselves, is the story of this spectacular test!

Chuck Stevenson   Clay Smith   Johnny Mantz   Bill Stroppe   Chuck Daigh   Walt Faulkner

Representative of the stock car winner—a Lincoln at the take-off!

And one of the competition Mercedes at the starting line!

day 1

Tuxtla Gutierrez to Oaxaca 330 Miles

We knew the start would be tough...

"There's only one way to win a race—and that's to dig in from the start," says Chuck Stevenson, this year's AAA national champion and driver of the first place Lincoln.

"With starting positions based on entry dates, our Lincolns, first of the 1953 production run, were entered late. So we had to pass to win, and the first day's 330-mile lap was all up-hill running. Our time would have to be made up in bursts on short straightaways, in superior handling qualities on the curves.

"With Lincoln's new 205-horsepower engine we had that important first advantage. That's why, when we got to Oaxaca, Lincolns driven by Walt Faulkner, Capt. Bob Korf, Johnny Mantz, Duane Carter, and myself held five out of six places. With that engine, and Lincoln's ball-joint front suspension, we had the advantages. Only Pat Kirkwood in a competitive second place car cut into the Lincoln record. The time allowed was 6 hours and 15 minutes. Saving our punch for later heats, we did it in 4 hours, 10 minutes, 49 seconds . . ."

Clay Smith, co-driver, points out 4-barrel carburetor of the 1953 Lincoln to driver Chuck Stevenson. Housed inside the air filter normally, this precision fuel meterer is designed to get maximum efficiency out of the engine's turbulence-type, kidney-shaped combustion chambers. "Given controlled turbulence, gas does more work more efficiently," says the nation's most widely recognized race and production car mechanic.

day 2

axaca to
exico City
36 Miles

10,000 feet mean a lot of curves...

"You begin to feel real affection for a factory-new car when you're saw-toothing mountain country, cutting corners in the toughest run in the driving business. Take Lincoln's very special front suspension. The ball-joint design—in which the wheel both turns and takes the up and down movement of springing on a single axis—let us wheel into the turns at a rate that picked up extra seconds. No heeling or digging—and this car lets us "drift," or slide, just like a race car. In this car and in this mountain country you drive by speedometer, because both steering and control are so easy you can hold her right to a schedule. At the mid-day break, Clay put in a new set of carburetor needles to compensate for the climb—as high as 10,483 feet. And all along we were taking advantage of the Ferrari Team's road markings, painted right on the pavement: for a hill with straightaway, two straight lines. For a single curve, a circle. And for a sharp curve, a circle with a slash on the side we'd be heading into. And now it's really downhill all the way into Mexico City!"

Here, Johnny Mantz and Bill Stroppe take their car, No. 122, around the new acceleration run return loop at the famous Ford Motor Company Dearborn test track. At high speed they show the road-holding stability of Lincoln's ball-joint suspension, available on no other American production car. Note how here, as in the picture above, all four wheels hold the road—and the almost complete absence of "heeling" or "scooping."

59

day 3

Mexico City
to Durango
612 Miles

Like riding a steel bridge across Mexico...

"This morning we roared away from the world's most enthusiastic crowd. This is the country's national holiday weekend and along with this race, it's the biggest celebration on earth. The cars are really breaking-in smoothly now and we've got our biggest run of the race ahead, through Leon and then over the hills to Durango. The straight stretches last a little longer here, and you get the full enjoyment of this car's balance. Wonderful the way they hold the road. Faulkner hit a bump yesterday and took right off—all wheels at least four feet off the ground. Just sailed through space 'til her hood nosed down slightly, he said, and off he rolled, almost without a skip. Came in for the nicest landing! Balance is the key to performance like that. The jounce didn't loosen a rivet, and the wheel-alignment held its factory set. Ball-joint suspension and one of the huskiest frames in the business—a real steel bridge to cross Mexico on. We're really making time, now, and our cumulative total puts us in third place, with the other three Lincolns just ahead and behind us!"

The chassis of your car is the steel bridge you ride on. Chuck Stevenson and Clay Smith demonstrate Lincoln's chassis ruggedness and the car's inherent balance by taking the test track "railroad crossing" at a speed that would nose-roll some high performance cars. Where No. 129's rear wheels touch, only tire burns will show. The car itself will be a mile down the track in less than 60 seconds!

day 4

urango to
hihuahua
37 Miles

Now we began to cruise the straights...

"The roads are really beginning to straighten out, and now our reserve power begins to show. Some who pressed at the earlier stages will burn out today. (We learned later that four more cars dropped out.) We're beginning to move up on others in our division, riding their tail and letting their windstream give us a "tow" and our engine a let-up. You do this for miles almost bumper to bumper, then swing around and begin the fight to pass. (It calls for steel-nerved driving, and you'd better know every inch of the road ahead of that other car. One 'follower' behind another of our Lincolns remembered a right-angle turn too late to 'drift'!) The minute you break away, your engine really has to work. Along the way we pass by four foreign jobs, but it's tough for one of these Lincolns to pass another Lincoln—unless you've got miles of straightaway to press on. It's no longer just chance that's giving this new Lincoln the name it's winning for itself. It takes more than driving skill to bring four of us in together each night—within minutes and seconds of each other."

In cross-country driving as much as in traffic, full visibility is both a comfort and a safety bonus. This view of a Mexican Race Lincoln windshield demonstrates the point: top is flat and extends clear to receding corner posts. No need to duck around curving blind spot. Bottom is low—to take advantage of extra visibility sloping hood affords. Mirror, tachometer are specially placed for race.

day 5

Chihuahua to
Juarez
230 miles

Winning seemed almost automatic...

"We've covered 1,708 miles of road, grades, peaks and plains since Wednesday, and now we're heading for the pay-off. The road is really straight now, and we're set to take it in stride. And we do—covering the distance in a shade under two hours—by just 10 seconds. We keep our windows up, to get maximum smoothness as we cut through the air. The ride is still easy, and we're not really tired, except maybe Clay is almost glad that tonight doesn't call for another sharp tune-up. Now all we hear is the drum of the floor—no floor pads to collect dust that spins and blinds on a run like this. The engine temperature keeps its unvarying reading, and now, up ahead you can see the finish mile crowds. Well, we've done it—swept the field, even passing all but six of the sports cars—and all the way from Mexico City on it's been Lincoln —just Lincolns, bunched by second and minute around us. Quite a record, and because of their even output, the sharpest driving I've ever had to do to win. You ought to try this car yourself—for driving, like winning, is almost automatic!"

The mile-eating cruising pace, with every driving move and every engine revolution in perfect harmony, is a real "money" ride, say Lincoln's Mexican Pan-American Race drivers. And Lincoln's integrated design—steering, springing, balance and power—all add up to road-ability that relaxes. Here a Mexican Race Lincoln rides out the Ford Motor Company high speed track in record time— just well broken in by its rugged run.

the finish:

The third running of the Mexican Pan-American race fulfilled the traditions already gathering about this spectacular classic. This year, though, due to its two divisions, there were two first place winners—Chuck Stevenson, in a Lincoln in the Stock Car Class, and Karl Kling, in a Mercedes 300-SL—in the Sports Class. Of 27 sports cars, 10 finished.

Of the field of stock cars, 29 finished. Of the Lincolns entered, the four winners stayed together all the way—a feat so singular that officials were asked to re-check the cars!

Winner Chuck Stevenson won first place with a total time of 21 hours, 15 minutes, 38 seconds. Only 31 seconds behind came Johnny Mantz with Walt Faulkner finishing third, less than 5 minutes behind the leader.

After the race, at the request of the drivers' representative, race officials again made a complete technical check of the winning Lincolns to reaffirm their qualification as stock cars—so unique was their sweeping win.

1st — No. 129—the first Lincoln—crosses the finish line a winner.

2nd — ...followed 31 seconds later by car 122—another Lincoln.

3rd — 4 minutes, 18 seconds later, the third place Lincoln crosses the line. Then...

4th — ...in 4 minutes, 42 seconds, fourth Lincoln finishes—20 minutes ahead of next car.

...the winners came in driving Lincolns

**TEST CAR AT A GLANCE—
'54 Lincoln Capri
With Power Steering
and Power Brakes**

REAR WHEEL HORSEPOWER 
(Determined on Clayton chassis dynamometer; all tests are made under full load, which is similar to climbing a hill at full throttle. Observed hp figures not corrected to standard atmospheric conditions)
46 road hp @ 1200 rpm and 31 mph
56 road hp @ 1500 rpm and 37 mph
82 road hp @ 2000 rpm and 54 mph
112 road hp @ 3100 rpm and 83 mph

SPEEDOMETER ERROR
Car speedometer read 32 @ true 30 mph, 51 @ true 45 mph, 68 @ true 60 mph

(In acceleration and fuel economy, the car is rated against the average of '53 test cars in its own class. All other items are rated against the average of all '53 test cars.)

ACCELERATION 
(In seconds; checked with fifth wheel and electric speedometer)
Standing start ¼-mile (76 mph) 18.4, 0-30 4.1, 0-60 12.3, 10-30 3.5, 30-50 5.9, 50-80 15.0
RATING: EXCELLENT

TOP SPEED 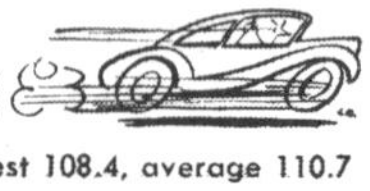
(In miles per hour; clocked over surveyed ¼-mile)
Fastest run 113.8, slowest 108.4, average 110.7

FUEL CONSUMPTION 
(In miles per gallon; checked with fuel flow-meter, fifth wheel, and electric speedometer. Mobilgas Special used)
D-3
15.7 @ steady 30 mph, 14.1 @ steady 45
D-4
19.8 @ steady 30 mph, 17.9 @ steady 45, 14.6 @ steady 60, 12.3 @ steady 75, 13.7 in simulated traffic over measured course, 14.4 overall average for 933 miles
RATING: AVERAGE

STOPPING DISTANCE 
(To the nearest foot; checked with electrically actuated detonator)
42 @ 30 mph, 92 @ 45 mph, 167 @ 60 mph
RATING: EXCELLENT

Handling is tops, brakes are better than ever—an all-around good, big car, as shown in this brief report of the...

'54 Lincoln

By Walt Woron
AN MT RESEARCH ROAD TEST REPORT

HAS THERE EVER been a time when you've gone to a movie and as the scenes unfolded in front of your eyes, you've said to yourself, "Haven't I seen this before?" It's happened to me, but never before on a road test.

I should have known this would happen though, because I didn't really expect anything to be radically different when we tested the '54 Lincoln; it's mighty hard improving on something that's already so good. Sure, it can be improved—and will be—but it's refreshing to see a manufacturer produce a good product and then stick with it for more than one year without changes. By next year, the pressure of that segment of the car-buying public that wants change for change's sake will undoubtedly force radical differences which are only whispers now. In the meantime, the '54 Lincoln is more than good enough, as over 10,400 new owners (as of May 31) will testify.

During the road test I was asked by many people, "Just how good *is* the Lincoln?" So many had seen reams of publicity on this car as a result of its 1-2-3-4 win in the Mexican Road Race that they couldn't quite believe it to be that good. My answer, this year again as last: "It's an all-around good, big car. Handles sweetly. Has lots of punch at all speeds. Rides comfortably. Brakes are much better than last year. Has absolutely no major faults, unless you expect the greatest fuel economy in a big car and are concerned with getting the highest trade-in value."

As for the details, the car is virtually the same—and I mean in all respects, looks and performance. It's been thoroughly covered in former issues of MT (Mar. '53 and Jan. '54), both test-wise and opinion-wise. Nevertheless, in our effort to give you complete data on all cars, and since we're constantly improving our testing techniques, we thought it advisable to test those features that should show improvement over the previous model. Where performance would not vary we give the figures for the '53 Lincoln.

The one big change in the '54 Lincoln is its brakes. The use of brake drums that are larger (by one inch) gives 18 more square inches of lining area (up 10 per cent over '53). More important, stopping distances have decreased appreciably, from an overall average (of stops from 30, 45, and 60 mph) of 136 feet in '53 to 101 feet this year. In this respect the car has been brought from below the norm to above norm (when compared to all '53 test cars). One fault was apparent, however. After repeated brake applications during the high speed acceleration checks,

Not the Mexican Road Race, but MT's test car showing off its roadability in Tom Medley's shots. Rear wheel slide

the brakes faded severely. This did not occur during the actual braking tests, nor did it happen during any of the rest of our hard driving.

Roadability is still tops, as is the car's ease of driving. Going cross-country, whether you're cruising along happily at moderate speeds, pulling around a line of poking trucks, or winding over hill and dale, you'll be relaxed behind the wheel, particularly if it has power steering, as our test car had. Anytime you have to call on the power that's there, you'll know it as soon as you punch the throttle. And when you *are* booting it, you'll be able to go into turns faster and have more positive control than in any other car we've tested this year (even with power steering, which retains more feel in a Lincoln than in any other car). It's set up for oversteer; the rear end can be made to break loose on sharp turns taken at 50 mph or above and you can then slide through or pull it through with control by punching the throttle.

While you're driving this way (which you're liable to do because the Lincoln almost begs to be driven hard) your passengers won't complain too loudly. The car leans and they'll shift in their seats, but aside from the fact that the armrest doesn't provide a good anchor (it's solid and allows only a fingertip grip) they won't be cussing you for driving too fast.

Last year we complained of tire squeal. This year it's practically gone; the tires have built-in squeal suppressors in the form of small rubber buttons between the ribs. Largely because of the broad-based tires the car still whips aside badly on streetcar tracks, lengthwise grooves, or tarstrips.

Last year we rated the Lincoln's acceleration as above average. This year we got a definite improvement by altering our shifting technique. In '53 we used D-3 alone, which starts the car in low gear, takes it rapidly through second and into third. What we did this year was to start in D-3, wait until it automatically shifted into second, push the selector to Lo range to keep it in second gear, peak it in that gear, then manually shift out to D-3 (third gear). The use of the lower (and quicker) gears improved our acceleration up to 60, but beyond that there was no improvement over the method used in '53.

If you're driving the car ordinarily, you don't have to shift this way. You'll probably put it in D-4 and keep it there, unless you want to use D-3 for more engine braking or for manual control while passing, and you'll want to use Lo for pulling a steep grade. During automatic shifts, the Hydra-Matic acted smooth enough except for an occasional "clunk" or a heavy lurch at low speeds under full throttle acceleration when it shifted from second to third or third to second.

For fuel consumption (Mobilgas Special was used) we have one additional figure not given last year. That's the overall tank mileage we got from the time we drove away from the Los Angeles factory branch until we returned the car 1319 miles later (aside from acceleration and braking checks).

That's about it except for a few minor trim changes like gold background instead of black behind the instrument panel numbers and letters (incidentally, they're less legible), a more distinct bumper-grille combination (a practical item), larger capacity oil passages in the hydraulic lifters (to minimize the noisy operation following a period of stopping), and new color options. Another practical item, and one we're still raving about, is the Lincoln sedan's use of plastic for door panels and headlining covering; it's an added bonus of "livability." About it, that is, except for our consensus that the '54 Lincoln's size is handy, forward vision is great (even without a wrap-around windshield), type and workmanship of trim is above average, ride is both comfortable and firm, while its roadability is tops. It's a lot of money, but a lot of car. **CONTINUED ON PAGE 91**

on dirt road was corrected easily by nudging responsive V8 into action. Heel-over, body sway are at a minimum

65

with the Lincolns in Mexico

Nothing less than perfection can satisfy Lincoln's factory team when the question is "Ready for the Race?"

By Pete Molson

MEMO from Walt Woron
To: Pete Molson, Motor Trend
c/o Robert W. LaMontagne, Jr.
Comite Nacional de la Carrera
Panamericana
Av. Cuauhtemoc No. 242
Mexico 7, D.F.

Dear Pete:

First of all, I hope your trip to Mexico City was enjoyable and not too tiring. We didn't get much chance to discuss slant on this story, except in that phone conversation, so here's a reiteration:

1. That MT would like to see more factory-team competition in the road race, for the good of all companies, and particularly for the good of the motoring public.
2. That behind-the-scenes cooperation is not enough. It takes factory cooperation like that given by Lincoln to make it really count.
3. That just because Lincoln has won three consecutive years is not reason for other companies to become discouraged, or to feel that Lincoln has it all "wrapped up." They would

Chart on Bill Stroppe's office in Long Beach shows all trip details, plan for service strip at end of each day's leg

(officially) like to have some competition.

4. That it doesn't take unlimited resources to run, but that it takes a percentage of what you want *out of it*, regardless of what that amount is. It's an investment in publicity, in knowledge, in experience.
5. That Lincoln isn't the *only* organization that can prepare their cars the way they do, but if anyone is to win (or even to run sensibly) they can take a few pages out of Lincoln's book.
6. That what it takes is months-long preparation of the cars, of the pit crews, of the drivers, of preparing for any eventuality, of training over the road, plus luck—and lots of it. Best of luck. I envy you.

Cordially,
Walt

MEMO from Pete Molson
To: Walt Woron

This story begins way back, not on the 19th of November in Tuxtla Gutierrez, nor on the first of that month as five trucks and, on the backs of two of them, nine race cars (seven Lincoln hardtops and two Ford coupes) crossed the border from El Paso, Tex., to Ciudad Juarez, in Mexico's State of Chihuahua. It began in the minds of the big men at Lincoln-Mercury, and as much as 20 years ago in a way of life chosen by four dozen rugged men for whom racing *is* life. It began in a modest-appearing but superlatively equipped shop in Long Beach, Calif., whose staff had long been known to the Ford Motor Co.'s neighboring assembly plant and to its parent Detroit office as the best trouble-shooters in the business.

The shop is one you would never notice, a low, typical Southern California small business building, looking a little like a trim suburban house. Its owners are Bill Stroppe and the late Clay Smith. Behind its yard gate and the friendly "Please! No visitors until December 1—time is short!" the organized hassle stretched farther into each night, both helped and hindered by the knowledge that Lincoln had won twice before and that a lot of people were waiting for it to win again. Most of the public knew little of what winning a race involves, but that actually meant little: the thing to do was to win, not explain why they hadn't. Paradoxically, where painstaking preparation was the largest ingredient in victory, it would carry no weight whatsoever without an outright win.

Let's take time out for some of the theory behind all the seemingly frenetic (but actually cool and collected) activity in Long Beach. The name of Lincoln has acquired that winning sound through a series of hard-headed steps. The first was one of the legacies of Henry Ford. As any amateur rich man knows, you can't buy shares in Ford (in 1954 there was a brief flurry of rumors to the effect that you soon could, but it remained just rumors). This single fact of life has written more chapters in the Decline and Rise of the Ford Empire than there are in the accounts of half a dozen stodgy stock companies put together. A stock company is obliged to show a profit, and the only way it can do so is by selling the public what it wants. But when Ford wants to go on making Model Ts 'til doomsday, or wants to introduce a V8 at a singularly unpromising point in U.S. economic history, it's a free agent. And, in plain talk, if it wants to spend a lot of money trying to win a still-obscure, 2000-mile race against frightening opponents and over appalling terrain, it doesn't have to answer to anyone. So Lincoln, incredible as the fact may seem to the factory drivers of Lancia, Mercedes, or Alfa, had the only true factory team from our reasonably prosperous country.

To be sure, other U.S. factories helped the entrants whose cars bore their names. Test and research facilities were at the disposal of loyal drivers in a sort of lend-

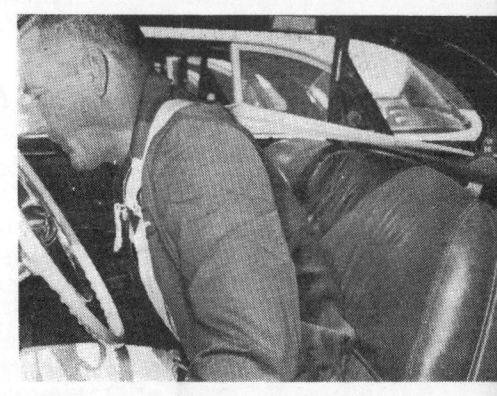

Chuck Stevenson shows how only slow movements will make safety strap yield

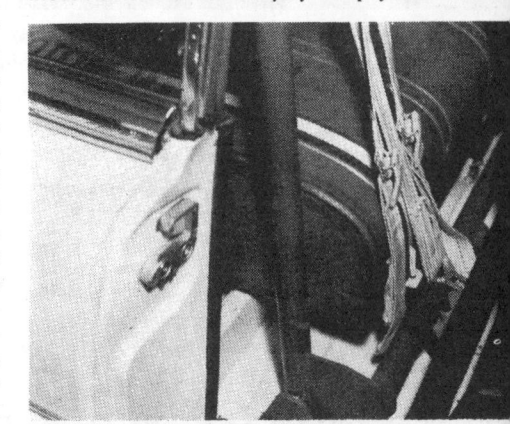

Driver attaches to belt, which attaches to roll bar, which bolts to the floor

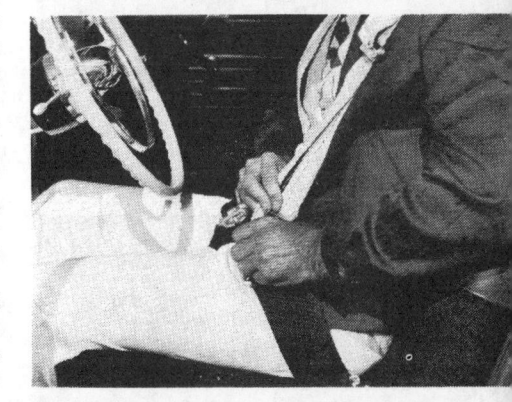

Chuck flips one buckle which obligingly releases both belts in case of emergency

Photos by Tom Medley and Pete Molson

Here's an impromptu view of the Lincoln crew as they prepare for a practice tire change. Note the electric lug wrenches

Here pilot and co-pilot begin the job, aided by a power jack which works off an air tank visible at the lower right

Leaving a wake of dust, they take off exactly one minute, 48 seconds later, with both the left and right tires changed!

WITH THE LINCOLNS IN MEXICO continued

Caravan consisting of five trucks (two of them carrying seven Lincolns and two Fords between them) pulled into this Mobilgas station in Long Beach before leaving

Neat stacks of toolboxes are always kept on hand by crew. Color of service kits corresponds to color of its assigned car

An engine dynamometer in the Long Beach shop. Don Francisco, formerly of Hot Rod Magazine, checks it

Two-way radio equipment in the back of Chuck Daigh's Lincoln made it possible for the caravan trucks to communicate

Bill Babbitt welds a hood-latch with portable equipment taken along in truck. Note the bottled drinking water in rear

lease arrangement, almost as though the factories would rather not win. (Some bitter would-be winners preferred to call it "too little and too late.") What it takes is just what Lincoln offered: an unstinting, flexible contract with a thoroughly proved outfit to see that the cars are in the best possible trim on race day, that every modification possibility is explored, that not a nut is left unturned to make each car a potential winner. It requires a generous hand with many thousands of dollars, though the sum was not a large one in terms of investments. It takes money available for *everything* that's needed, and in as mercuric an art as racing, a cold *cerveza* (beer) for the guy who has just changed a truck tire can be just as vital as the proper-sized end wrench. It takes a thoroughgoing awareness that you have to get ready to win: luck, mumbo jumbo, or prayer won't make it so.

Now, Lincoln, aware of what can happen on both sides of the ledger, takes a calculated risk. Publicity-wise, the '54 win was tremendous. The possible loss of the '54 race had to be borne in mind, for the U.S. car-buying public is quick to forget a favorite. But Lincoln had no choice in the matter; failure to enter would have been worse than losing, and could not be considered. Actually, Lincoln wants, and will continue to want, close competition. The fact that Chrysler won in '51, came close last year, and was serious (if not serious enough) about its '54 entries made it the biggest and most threatening of bogeys. At no time was its presence treated lightly, and it was an eye-opener to see these experienced drivers and mechanics carefully estimating their enemy.

This estimate began with the anonymous purchase of a new Chrysler, which Lincoln carefully tore down and inspected, bit by bit. Frame, suspension, Power-Flite unit, all came under the magnifying glass and made possible an estimate of the car's behavior in each contingency up the 2000-mile length of the course. Can a Chrysler corner as well as a Lincoln? "Not in stock form" is a quick and easy answer; but what happens when permissible safety modifications are added? I watched one of the Lincolns scorch the fantastic road south of Mexico City; I was a very few feet in front of it (in Johnny Mantz's Lincoln, No. 104) and it leaned not at all when going through sharp curves at 60 mph. MOTOR TREND's readers know that stock Lincolns corner among the best U.S. cars, but only a fanatical devotee would claim that they go through such hair-raising spots *with no lean at all*. I didn't see the Chryslers in the uneasy (for me) performance, but it's a cinch they behaved differently from their boulevard counterparts.

Let's consider, then, what could happen. A Chrysler is ahead of a Lincoln, say, going into a turn. The Lincoln driver knows approximately what the Chrysler can deliver from his study of its transmission and from its roadability, and of course he knows his own car even better. If the Chrysler swings wide on a turn, an experienced Lincoln driver can charge through the hole and be out in front while someone else would be wondering just what to expect from the car ahead.

Let's go into Clay's and Bill's shop and see what makes it work. For the cars' heart, there's an engine dynamometer room, no small investment in itself. Making sure that a car is in tip-top health is tricky with a dyno, but the important point is that it can be done with it, and not as well without it. Almost any man, driver or mechanic, on Lincoln's team can cock an ear from the front seat of a car and hear a slight unevenness that you or I would never detect, but the dyno can go farther than this: from long experience, testers like Don Francisco (formerly of MOTOR TREND's sister magazine, *Hot Rod*) know what kind of dyno reading they should get from a perfectly balanced (though stock) engine, and they're satisfied with nothing less. When an engine is running right, and only then, it's dropped back into the chassis—to do its part in pulling the Lincoln team to its scheduled victory. (A dynoed Lincoln engine even pulled one of the race car transports, insuring its efficient performance in getting the team to the starting line without mishap or delay.)

Out in back of the office, behind an unimpressive board fence, was the tire stop, set up for practice exactly as it would be at the super-efficient spot near Tehuantepec, south of Mexico City. It was Saturday afternoon when I first saw this, but a purely workaday feeling filled the air. The drivers took turns roaring around an improvised dirt track and sliding in for a stop on metal tracks, buried flush with the ground. (Once Bill Vukovich, in an excess of enthusiasm, dug the whole assembly from the ground and it had to be installed again.) Out leaped the co-pilot (while the driver settled the front tires into their grooves) and flipped the air valve to operate the jacks, which had previously been centered precisely under frame crossmembers. A replacement tire, complete with balanced wheel, was propped up near each corner of the car; electric lug wrenches lay handy to each wheel; each new wheel had a lug hole daubed with white paint corresponding to a lug similarly marked. (Balance was exact, and not to be marred by doubt as to where the lug went.) Round and round the crews went, narrowing their time, spotting the danger points that could cause trouble. "One minute, 58 seconds!" the timekeeper called, and while I watched, subsequent crews cut this by 11 seconds, and later, practice brought it down still more.

Next (and all in the same dusty rear lot) came a performance like a "Medic" show; in this case, a Lincoln was the supposedly ailing patient. Each car was assigned a characteristic color (orange, purple, or whatever) though the body color of all the team cars remained white. In the dust was unrolled a snowy cloth; on it went the tool box for the car in question (it, too, was "color-keyed") and from it, mechanics took the tools, to be spread out on the cloth. In this dramatic manner, each car would be serviced at the end of each of the five days' run. On the actual caravan from Long Beach down to the starting line, the crews would inspect the service areas so they could know every move during the race; and before they even saw the actual strips, detailed plans of all of them—Oaxaca, Mexico City, Durango, Chihuahua, and the extra tire shop—hung on the wall in Bill Stroppe's office for study.

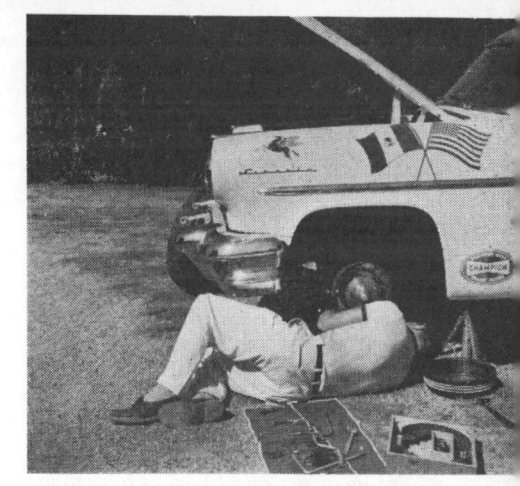

In Oaxaca, mechanic works on shock absorber and brake. As usual, a neat, operating-room-like condition prevails

Scene: Oaxaca, after practice trip from Mexico City which showed up snags in suspension, brakes, shock absorbers

Radio mike and auxiliary instruments in Chuck Daigh's sedan used for testing the course and general Lincoln behavior

Interior of a race car shows footrests for driver and co-pilot. Note big tachometer, pressure and vacuum gauges

Voilà! Gordon Smith cooks on a butane stove set on tailgate of the chuck wagon. Max Ellis is the chief bacon tester

With the Lincolns in Mexico

Bewildered, I wandered back inside. Final touches were being put on the two-way GE radio installations in Chuck Daigh's test Lincoln and in the caravan's trucks: no time was available for the caravaners to get separated and lose valuable time assembling. (Thus, from Long Beach to Mexico City, and to a lesser extent from there south to Tuxtla, any difficulties were quickly known to everyone in the caravan, no matter how wild and isolated the road or how dark the night.)

Rooms piled high with carefully painted wheels, each keyed to a car and balanced, were chalked with legends like "LR OXA 108." The meaning of this cryptic notation was that the tire would be unloaded from the parts truck at the Oaxaca service strip; that it was balanced for the left rear wheel; and that it would go on Bill Vukovich's car (No. 108).

Bill Stroppe and Gordon Smith were off to Mexico weeks before the actual caravan set out on its way. Aided by complete films of the '53 race, taken from the co-pilot's side of the front seat (they have the movie "Rear Window" beat a mile for tense watching), these two charted every foot of the course from Tuxtla to Juarez. No curve would take a Lincoln driver unaware, no bridge would be narrower than he thought. It's all in every car, on compact charts mounted on rollers in a neat case. Every co-pilot can call off what's coming up, safe speed at every bump, things to watch out for. Forewarned is forearmed. Here's one of the spots where economy—specifically applied to the number and quality of men a manufacturer hires for his team—just can't pay off. If Stroppe had had to go to Mexico and leave work undone in Long Beach, results would really have suffered; but there was a full crew in Long Beach, and preparation continued while he was away.

Actual departure of the truck caravan was delayed a couple of days for truck-wheel balancing, repositioning of spare parts on the compartmented service trucks, and consideration of *diet.* If this last sounds strange, remember that travel can mean distressing, though minor, upsets to the best of us; taking chances that a driver or even a key pit crew man will come down with the collywobbles is just not playing it smart. So dozens of huge demijohns of pure spring water went on top of the parts truck; an enormous refrigerator had its own gasoline engine, and contained steaks galore (sirloin tips did for stew!). The latest type of canned-but-fresh-tasting milk, not obtainable at retail, was ordered. Candy bars and cigars appeared in gargantuan boxes. (Man cannot live by bread alone.) When the time came to leave, the caravan was entirely self-sufficient with the exception of gasoline and, of course, sleeping space.

A result of these careful plans was that the trip across our own Southwest, and from El Paso down to Mexico City, was definitely on the dull side so far as race-business went. It was a useful breathing space for many of the drivers and crew (but of course hard work for the truck drivers; we live and learn, and no one had better say, "Oh, he's just a truck driver!" to me again).

A few days in Mexico City sufficed for finishing up pre-testing jobs, checking brakes, etc., and one quiet afternoon we departed for Oaxaca, with me sandwiched between Johnny Mantz and Bill Stroppe, in the car which they hoped to bring in first. At first I thought that I had somehow slept over 'til race day by mistake. When you ride with a professional race driver for the first time, it's bound to be a thrill; when it's on a practice try at one of the toughest legs of the Panamericana, yipe!

To begin with, the two superb safety belts and shoulder harnesses, cannily planned to allow free movement when driving, but virtually none in an emergency, were useless because there were three of us in the seat.

Following too close to another car has become a major traffic offense in many states, and with reason: most drivers, certainly including myself, just aren't up to the sudden demands that bumper-riding can make on them. Yet three of the race Lincolns (including Mantz's) covered the 300-odd miles *a few feet apart,* and with a feeling of security which grew with each bizarre curve. The reason was compounded, in equal parts, of the cars' safety features and the drivers' ability. *No* driver could have driven a totally unaltered car in such a way and lived to read this; and no car, no matter how good, could have sat so solidly on the road in less expert hands. My ride was anything but comfortable physically (a plugged gas tank vent and the multitudinous curves combined to make me carsick, for the first time in my life and to the amusement of the drivers), yet before we pulled into Oaxaca (in about six hours, to the astonishment of the motel owner) I was nodding and napping along with Stroppe, though we both sat bolt upright.

Sliding into a corner is one thing, I found, on MT's broad and unobstructed test strips, and quite another on a hairpin curve at 60 mph, with a rocky bank on one side, a hundred-foot cliff on the other, and four bicyclists approaching in the center of the road (all practicing, of course, was done during "normal" conditions— with such hazards as the aforementioned bicycles, burros asleep in the road, and various pedestrians, all of whom seemed to be under four or over 70). The racing tires screamed softly—they didn't sound half so bad as some stock sedans—Mantz flipped the Hydra down into D-3, and we'd be off again without a lost second, and with, so far as I could see, no loss of control.

Huge chuckholes abounded in short stretches of the road, and though some resulted in a sort of Mantz version of broken-field running, others were so thick that there was nothing to do but plow through them with sheer brute force. I must have made quite a face as we approached the first of these compulsory minefields, for I noticed Johnny grinning at me just before we hit. But instead of it being all over for us, the Lincoln had murmured "dlump" and continued unperturbed. I don't think the hood even dipped. By race time, the crews who were filling in the holes and shoring up rickety bridges will have completed their job, and these hazards will not exist. But the crowd control—probably better here than on any comparable race in the world, if there *were* any comparable one, that is— can never do a perfect job, and time spent now in dodging cows and such always pays off in sharpened alertness. The only thing to dodge may be another car, but I don't need to tell you that that calls for pretty good judgment, too.

As for straightaways, they were few and short south of Mexico City. Chuck Daigh had thoroughly covered the area north to the border, more or less playing tag with our lumbering truck caravan. When Johnny got No. 104 up to an indicated 90 on its more-than-usually-accurate speedometer, it was nothing at all (at least to me; I guess I was blasé at that point). And of course, the last few hours' practice were run in the dark, a circumstance which slowed down Mantz not at all.

So it was a strange ride; their experiences on it were evidently unsatisfying to all the drivers, though this seemed incredible to me with such perfect-handling cars. On the morning after we arrived at Oaxaca, the crew had virtually torn all the cars to bits and laid them out on the grass. Suspension was the chief beef, and to improve it, shocks, springs, tires, and wheels were discarded and replaced. Of course, 300 miles of that kind of driving can cause a lot of wear, and there's no point in starting a race—or even another practice leg, which the Oaxaca-Tuxtla dash southward would be—with equipment that's not up to snuff.

And that—though it took me two very full weeks just to watch it—was a minute fraction of all the work done to get Lincoln in winning shape. The men, key to satisfactory preparation, all have so much experience either in driving or as mechanics, or both, that it's staggering. Many come from the aircraft industry, whose standards of accuracy are second to none.

CONTINUED ON PAGE 91

AUTO REVIEW

The new Lincoln Capri is distinguished by a redesigned grille and front ornamentation. Both of Lincoln's models include an all new Turbo-Drive automatic transmission and 225 hp. power plant.

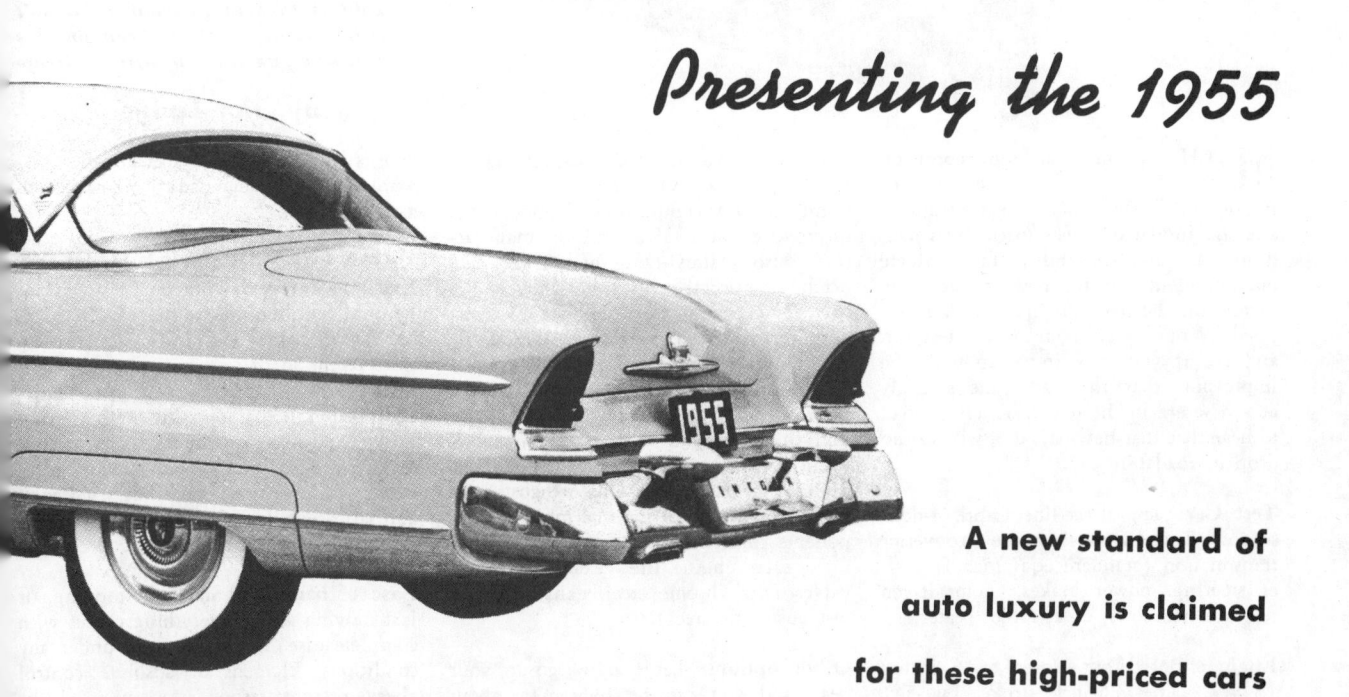

Presenting the 1955

A new standard of auto luxury is claimed for these high-priced cars

A modified rear fender and radically new tail lamp design are embodied in Lincoln's styling changes for 1955. The rear axle has been remodeled to increase tread to 60 inches and rear shock absorbers are at a more horizontal angle to enhance riding comfort.

'55 Lincoln Capri

MT RESEARCH road test

There's a feel of the future in this car that looks so like a '54

Lincoln is tops in roadability, doesn't sacrifice riding comfort. Even on this severe, skidding turn, lean is moderate

by Jim Lodge

MUCH OF this road test report on the '55 Lincoln is a synopsis of performance, the car's newest features, and the annual bouquets to the car's handling ease and roadability. But evaluate our comments on the new transmission, digest any between-the-lines feeling implied by our descriptions of the test car, and see if you, too, come up with the impression that this car hints strongly at a new era in the offing. You'll be glad to hear that that new day, tho, will see no drop in roadability.

Test Car: Top-of-the-line Capri, 4-door sedan. Turbo-Drive torque converter transmission (standard equipment), power steering, power brakes, motor-driven seat and windows, usual radio, heater.

Engine: Bored out closer to 4 inches (3.94), same 3.5-inch stroke, for 341 cubic inches displacement. Torque boost to 342 pounds-feet, rated brake horsepower now 225. Higher-lift camshaft for faster intake valve timing, giving decreased overlap for smoother idle, low-speed running. Engine breathing improved thru redesigned aircleaner, reworked 4-barrel carburetor; intake manifold passages larger by 10 per cent. Overhead-valve V8 has kidney-section combustion chambers (to improve exhaust-valve cooling; manufacturer also claims that they lower fuel octane requirement in spite of new 8.5 to 1 compression ratio). Higher oil reservoir, new oil-feed method to "practically eliminate" hydraulic tappet clicking when engine's cold. Lincoln engineers tell us: "If you hear 1 or 2 clicks on a cold morning, that's normal. If it goes on, something is wrong, and the car should be brought to an agency." Dual exhausts now standard. Almost all internal moving components constructed of new-type materials for greater engine life (example: intake valves now chrome steel, exhaust valves cast austenitic steel).

Other options: Let goodies go to your head and you can be lighter by about $1500. If you can afford it, you can have options of your choice, including all-leather upholstery, air conditioning (a cool $621), and, most novel extra, push-button chassis lubricator (described in last month's MT). There's no transmission option, but last year's 3.31 rear axle is still available on special order (no extra cost); it's mandatory on cars with air conditioning.

WHAT THE CAR IS LIKE TO DRIVE

Exit and entry, driving position, vision, general feel behind the wheel: All good, unchanged from '54. Vision excellent despite retention of non-wrap-around windshield. If you trade a Hydra-Matic-equipped Lincoln on this one, Turbo-Drive's PNDLR quadrant is all that'll seem strange, unless you include new-found transmission smoothness, silence, and apparent loss in pickup; against a stopwatch, loss isn't noticeable.

Ease of handling: Starting, stopping, or just moving along, everything comes with complete ease. No wheel fight under any conditions. Element of absolute control always present; responds to your touch the way a good, roadable car will. Compactness of car aids handling ease (it's a fairly large car, yet imparts feeling of being smaller than it is). Great psychological effect on those who hesitate to park anything bigger than the smallest member of the Ford family. Power steering easy, re-

In addition to the roadability and performance that again made it the Mexican Road Race winner, Lincoln retains handsome profile, but has a 7-inch-longer rear fender

Photos by Joe Moore

tains enough feel to remind you that you're guiding a 4470-pound missile. Still rates compliments as fun to drive.

Acceleration: Look at the specifications, Clayton chassis dynamometer results, then hazard a guess on performance. Horsepower, torque, displacement higher this year. Turbo-Drive less dynamic-feeling transmission than last year's Hydra-Matic. Rear-axle ratio down from 3.31 to 3.07. Checkmate. Differences between '54 and '55 cars no more than $1/10$-second in all checks but 30-50 and 50-80, where '55 holds slight edge. Despite higher-geared final drive, top speed is lower; car still maintains high cruising speed with ease. What about the new transmission? Akin to Mercomatic, Fordomatic in design, operation. One of the quietest, most effortless transmissions tested to date. Upshift normally goes unnoticed; shifting to LOW at higher speeds brings non-jarring deceleration in intermediate gear. Not as flexible as former transmission in mountains; plenty of coasting in DRIVE. Automatic low-gear starts possible thru full throttle detent. New planetary-geared torque converter unit has 2.1 to 1 maximum multiplication; low-gear start brings 2.40 gear into play, intermediate has 1.47 ratio. Full-throttle upshifts made automatically at 35 mph (low to intermediate), and 75 mph (into final drive range). Maximum kick-down points 20 mph (from intermediate into low), 71 mph (from drive into intermediate). Faster shift than many.

Braking: Nearly identical to last year's stops at all speeds tested. As noted on MT's data sheet, "this car could have

Like '54 car, Lincoln has good visibility, easy entry and exit, comfortable driving position. Seats, floor mats are luxurious

Like most upholstery and all paneling and trim in the test car, headlining is of washable, childproof plastic material

'55 LINCOLN CAPRI continued

been a record-breaker had it stopped in a straight line." Unpredictable "grab" to either side until brakes became hot enough to keep wheels from locking prematurely. If the mix from which different batches of brake lining are made varies only slightly, it can vary the coefficient of friction and cause uneven braking. Another possibility is that the anchor pins were riding too high on the test car, a condition for the dealer to correct. Fade resistance fairly low.

Roadability: Continues to be a favorite on MT's mountain test course. What you do with it depends on your intelligence and skill. Wherever it goes, it does it without front-end mushiness, without uncontrolled sliding, without protest. Drifts evenly, gives warning of every move. Turbo-Drive doesn't give immediate throttle response for power recovery, but car tracks accurately, breaks traction gradually enough for easy steering control. Even at top speed, moves like it was built as a high-speed machine. No wander, no front-end float.

Ride: Draws fine line between roadability and ride. In soft light cast by competitors, Lincoln is relatively stiff. Here, tho, seats add more than meets the eye; add to inherent comfort the car's excellent stability, and you come up with good riding by any standard. Rear tread now 60 inches (was formerly same as front, 58½). Rear shocks repositioned; passengers found car steady, bounces not as sharp as before. Good distance car.

WHAT THE CAR IS LIKE TO LIVE WITH

Unique in its class, retains extensive use of washable materials in headlining, door paneling, trim, upholstery. Some models, depending on outside color, make use of *Medic*-like interior — including white armrests, upholstery (and even white numerals on white instrument background). Only slash of color in this one is fascinating red sweep-second-hand on clock. The latter you're bound to like.

Riding in the front seat: Legroom, headroom good, plenty of space to stretch and relax. Seat is comfortable, armrest well proportioned. No serious projections in front of passengers; we look for Lincoln to go to glareproof, crash-pad-type panel in future.

Riding in the rear seat: Can't stress enough the niceties of not being thrown from side to side. Hip room, shoulder room not as great as in others, but difference isn't too noticeable to most passengers. Legroom adequate, plenty of space under front seat for toes. Headroom about average. Surroundings pleasant; center armrest nice feature, helps avoid sliding around on slippery, washable seats.

ECONOMY AND EASE OF MAINTENANCE

Fuel economy: Here, as in acceleration, slight difference from '54. New car bettered the old at 30, 45 mph. Overall tank mileage down, probably because of eagerness in trying out low-gear kick-down from standstill. Traffic mileage (with normal takeoffs) down only ½-gallon. Surprisingly good economy by comparison to '54, considering new-type transmission, added power. Credit compromising rear-axle ratio.

How did it hold up? Except for brake fade in acceleration runs, car had high resistance to test grind. New transmission functioned perfectly at end of 800-mile road test. Engine very smooth and quiet at idle, under acceleration, at cruising speeds. Body panels secure, no mysterious rattles anywhere during entire test.

Servicing: No styling or engineering gimmicks on warmed-over body, chassis or engine to complicate servicing, repairs. New chassis luber is simple.

SUMMING UP

Compiling results of a week's testing, we detect a note of quality not discussed at the climax of former Lincoln road tests. It's the quality of the smooth and quiet ride that's all-important in the prestige price class. Turbo-Drive will be prime mover in earning Lincoln a reputation for things other than its obvious handling and roadability traits. It reflects challenge to be projected by '56 models.

GENERAL SPECIFICATIONS

ENGINE: Ohv V8. Bore 3.94 in. Stroke 3.50 in. Stroke/bore ratio 0.888:1. Compression ratio 8.5 to 1. Displacement 341 cu. in. Advertised bhp 225 @ 4400 rpm. Bhp per cu. in. 0.660. Piston travel @ max. bhp 2566 ft. per min. Max. bmep 151.2 psi. Max. torque 342 lbs.-ft. @ 2500 rpm.

DRIVE SYSTEM: STANDARD transmission is Turbo-Drive, 3-element torque converter with planetary gears. RATIOS: Drive 1.47 x converter ratio and torque converter only (2.40, 1.47 and torque converter only, at full throttle thru downshift detent); Low 2.40 x converter ratio; Reverse 2.00 x converter ratio. Maximum converter ratio at stall 2.1 @ 1525-1725 rpm.

REAR-AXLE RATIOS: Standard 3.07, 3.31 optional or with air conditioning.

DIMENSIONS: Wheelbase 123 in. Tread 58.5 in. front, 60 in. rear. Wheelbase/tread ratio 2.01:1. Overall width 77.6 in. Overall length 215.6 in. Overall height (empty) 62.7 in. Turning diameter 45 ft. 7 in. Turns lock to lock 4½ (3¼ power steering). Test car weight 4470 lbs. Test car weight/bhp ratio 19.8:1. Weight distribution 56.4% front, 43.6% rear. Tire size 8.00 x 15 (tubeless).

PRICES: (Including suggested retail price at main factory, federal tax, and delivery and handling charges, but not freight.) LINCOLN 4-door sedan $3563, hardtop $3666. CAPRI 4-door sedan $3762, hardtop $3910, convertible $4072.

ACCESSORIES: Turbo-Drive standard, radio (with automatic antenna, rear-seat speaker) $131, heater $121, power steering $129, power brakes $43, power seat and power windows $178, air conditioning $621.

TEST CAR AT A GLANCE
'55 Lincoln Capri

REAR WHEEL HORSEPOWER
(Determined on Clayton chassis dynamometer. All tests are made under full load, which is similar to climbing a hill at full throttle. Observed hp figures not corrected to standard atmospheric conditions.)
74 road hp @ 2000 rpm and 35 mph
93 road hp @ 2500 rpm and 58 mph
Max. 106 road hp @ 3000 rpm and 74 mph

TOP SPEED
(In miles per hour over surveyed ¼-mile.)
Fastest 1-way run 104.2
Slowest 1-way run 104.0
Average of 4 runs 104.1

ACCELERATION
(In seconds, checked with 5th wheel and electric speedometer.)
Standing start ¼-mile (77 mph) 18.5
0-30 mph 4.2
0-60 mph 12.4
10-30 mph 3.4
30-50 mph 5.3
50-80 mph 13.7

SPEEDOMETER ERROR
(Checked with 5th wheel and electric speedometer.)
Car speedometer read 30 @ true 30 mph
47 @ true 45 mph
62 @ true 60 mph
78 @ true 75 mph
110 @ top speed

FUEL CONSUMPTION
(In miles per gallon; checked with fuel flowmeter, 5th wheel, and electric speedometer. Mobilgas Special used.)
Steady 30 mph 18.5
Steady 45 mph 17.1
Steady 60 mph 15.2
Steady 75 mph 12.2
Stop-and-go driving over measured course 12.9
Tank average for 808 miles 12.0

STOPPING DISTANCE
(To the nearest foot, checked with electrically actuated detonator.)
30 mph 41
45 mph 93
60 mph 167

Road Auto Age Test

Proving Grounds Report:

THUNDERBIRD LINCOLN and MERCURY

Here are the complete road tests and specifications for the three hottest cars Henry Ford ever built.

With the wheels cut hard to the right or left the Thunderbird can be driven at fairly high speed without rear wheel breakaway or excessive leaning.

IT'S A FUNNY THING how one hand washes the other in the automobile business. For over 20 years now Ford products, and especially their engines, have been the favorites among the hot-rod clan in this country. You can throw some sports-car enthusiasts into that group as well. These power-happy boys have been taking Ford, Mercury and Lincoln V-8 engines and hopping them up to what seemed to be fabulous horsepower and while the factory was mildly interested, they did very little in this direction themselves. But suddenly the entire country, in fact the world, has gone speed and horsepower crazy and what do the Ford people do? They turn around and hop the cars up at the factory and proceed to produce a line of the hottest cars ever seen on these shores. I know; I drove them.

Take the Thunderbird, for instance. We have all been hearing a lot about this car for the better part of a year and now, even though some of them have already been delivered, an air of mystery surrounds the whole conception of the vehicle. Is it or is it not a sports car? How fast is it? How does it handle? What kind of engine does it have? You've heard all these questions and more. Suppose we answer them.

Ford claims that the Thunderbird is a "personal" car, not a sports car. I, for one, agree with them. The definition of a true sports car is one that is fast, handles well, can be driven on regular roads, but can also be raced. The Thunderbird meets all of these qualifications except the last, not because Ford can't build a real sports car but because they never intended to. The Thunderbird has an engine that displaces about 4.9 liters, European style. This is not really a Ford engine, you know. It's a special Mercury job, very similar to the one in the new Montclair. Whatever it is, it pushes the car quite fast—up to nearly 120 mph—but a sports car with a 4.9-liter engine is traveling in pretty rough company. A 4.9-liter

Dashboard of the 2/3 seater is attractive, uncluttered. Instrumentation includes a tachometer which is, however, badly positioned, far off to the left. Trunk compartment is large for a "sports" car.

Author drove the car up to 100 mph with top down and windows up, found the slip stream to be very good. Optional cloth top folds behind seat.

Ferrari, for instance, can do 165 mph with ease and there are many lesser-engined unlimited cars than can top 140. Where does this put the 'bird as far as competition is concerned? Nowhere! It's not surprising, either. Most of these ultra-fast competition machines weigh under 2,500 lbs.; the Thunderbird weighs almost 3,200 or more than some low-priced sedans.

But don't get the idea that Ford's new baby is slow, because it's not. I took it out on the big banked speed track at the Ford test grounds at Dearborn, Michigan, and believe me, it was a thrill. After a few get-acquainted laps at 70 to 80 mph I began to let it have its head and it responded like an eager polo pony. There is one long straight on this track with a high bank at either end, and as I came off the first bank at the completion of one lap I put the accelerator to the floor with the speedometer just touching 70 mph. The 'bird leaped ahead instantly and began to gather speed smoothly but very, very fast. There was no wheel fight or wandering and as I went into the turn at the end of the straight I was doing just under 110 on the clock. On a shorter back stretch I was amazed to find that the car wanted to cruise at 90 mph but you'd never believe it was doing much over 60 unless you looked at the speedo. Even at 110 the car held rock-steady and there was plenty of cross wind that day in Dearborn.

When I had finished my runs on the high-speed track, I drove on into the tricky handling course that is laid out in the center of the Ford grounds. This is arranged in a series of flat but regular bends that get sharper at intervals. Starting with circles of a 300-ft. radius, they work their day down to a tight 150 ft. and they present a real test for any car's suspension and steering. I had a Ford engineer with me as I drove around and at first I took it pretty easy. The tubeless tires began to set up a steady howl at about 35 mph and I asked to have the air checked. We went back and found that two of the tires had barely 20 lbs. of pressure in them and so we promptly equalized all four wheels at 25 psi. Then we returned to the handling course. This time I didn't hear the squealing song until almost 45 mph and I began to get more confidence. I pushed faster and faster and found that, although the car does lean a bit at first, the angle doesn't increase much at speeds over 50 mph.

Thunderbird is powered—literally—by a V-8 engine that displaces 292 cu. in. That adds up to just about 4.9 liters, European style.

On the first time around the course all went well and when I hit the 300-ft.-radius turns on the second time around I was doing almost 65 mph. I kept up the same speed as the curves began to get sharper and I was starting to have a little more trouble with the steering. Then, just as I came into the 200-ft. turn, I hit a wet spot that may even have been a bit icy and suddenly the rear end was headed for the blue. I managed to correct enough to get the car in a straight line but it wouldn't stay on the road. We went off into the grass and mud, foolhardy tester and helpless engineer both tightlipped, but the car showed its stability by coming around slowly to face the road and never giving any indication that it wanted to turn over. This car, by the way, was equipped with a manual three-speed shift, and I was in second gear at the time, with my foot flat to the floor.

This momentary lapse proved one thing in a hurry:

Lincoln's front end has been simplified, much improved. Car's greatest distinction is that it is only Ford product with old-style windshield.

Handling on this big-but-compact automobile is even finer than last year. Here it is racing around a 150-ft. circle at better than 50 mph.

the car oversteers greatly. Now some people, including Ford's top engineers, consider this an advantage but I don't—not in a car as fast as this one that tempts you to come into curves at high speed. An understeer car will drift before it breaks away and will give you something of a warning of what's about to happen. When the car starts to drift you *know* you're going too fast and you can back off on the loud pedal. A car that oversteers, however, will lose its tail first and once this happens. you have an awful lot of correcting to do to keep the car on the road. If you don't make it sometime, it might be pretty embarrassing. A certain degree of oversteer is all right for normal passenger cars but I don't care for it on the Thunderbird.

Having thoroughly frightened the engineer on the handling course I asked him to help me with the acceleration tests. He agreed and we put the car through a series of breathtaking runs with the following results: Zero to 30 averaged 4.1 seconds. Zero-to-60 time was a fast 10.4 and from there to 80 mph took another 8.6 seconds. If you don't think these are fast times, try a few corrected-time runs with your own car.

The brakes on the Thunderbird are really fine. Fast "crash" stops from 60 mph produced little fade and not much dipping either and the car will stop in a straight line with your hands off the wheel.

What are my criticisms? Well, besides the steering, I don't like the wrap-around windshield which distorts badly at the edges, acting like a lens. Even the Ford engineers admitted this. Another minor point is that I think the soft top and not the plastic hard top should come as standard equipment. The car, stripped, sells for $2,695 including a four-way power seat. If you want a cloth top instead of the hard top it costs you $70 extra, but if you want both, the charge is $270. I think most people would rather have the hard top as an option. I know I would.

WHEN I climbed behind the wheel of the new Lincoln Capri I got still another thrill. Basically the same as last year's model but with some engine and chassis modifications, this car gives the impression of being expensive

Beautiful Lincoln Capri can hold its own style-wise and performance-wise with any car in America. In stock form it approaches 110 mph.

Here's what the tricky Ford handling course looks like through the windshield of a Lincoln at speed. Tight curves are left unbanked.

Push-button lubrication, operated from inside the car, is optional on 1955 Lincolns, Mercurys. Oil reservoir is under hood, as shown.

but not gaudy. It is the only automobile in the Ford line that does not have a wrap-around windshield and this pleased me, as you may have guessed it would. It does have a new wrap-around bumper, though, and a few other styling innovations that make for real class.

But the big question to me was "how does it drive?" So I took it around the same route that I had followed with the Thunderbird. Toward the end of my third lap on the speed track, as I neared that same last turn leading into the straight, I held the needle at 70 mph, just as I had done with the 'bird. Believe it or not, this car was actually faster than the other bomb had been from 70 to 100 mph. I got it up to an indicated 115 by the end of that straight and although the Lincoln won't match the Thunderbird for top speed, it will climb right up to almost 110 mph, and with no dead spots in acceleration. The answer, of course, is torque, which is a factor of just plain engine size. The big Lincoln V-8 puts out 225 hp this year and develops 332 ft./lbs. of torque at 2,500 rpm. The Thunderbird's torque rating is 286 ft./lbs. at that same 2,500 rpm. That's where the difference in dig comes from.

On the handling course the Lincoln showed me why it has won its class in the Mexican Road Race three years

Clever use of chromium on the Capri minimizes the "slab-sided" effect, makes the car appear more graceful. Wrap-around bumper is new.

Two-door Mercury Custom is one of the least expensive cars in the line. Power is supplied by a new 188-hp V-8 engine that pushes it over 100 mph.

79

Mercury performed well on the speed track, got up over 100 mph with ease. Banked turns could be negotiated at 85 mph without any strain.

Dashboard has all instruments grouped in a V-shaped panel in front of the driver. The big speedometer goes up to an indicated 120 mph.

running. This is a superb handling automobile—beautifully balanced with fine, accurate steering and the best front suspension of any American car I have ever driven. On the small 150-ft. circle the Lincoln went around under perfect control at speeds up to 55 mph. This kind of stability and feel, plus all the luxury Lincoln has to offer, is going to be mighty hard to beat.

An interesting fact that most people don't realize about the Mexican thing incidentally, is that this past race was won by 1954 Lincolns, not the brand new models. Race rules state that the year of the cars in competition must correspond with the year the race is run. This is to make sure that one manufacturer doesn't get an unfair advantage by jumping the gun with his new models. Just thought you'd like to know.

Getting on with the test, I went into the usual acceleration runs but since it had rained in the meantime I would prefer not to give my exact figures since they would be neither accurate nor fair. I will say, though, that the 1955 Lincoln Capri will

Interiors are brightly colored and beautifully designed. A combination of heavy plastic and fine fabrics should prove quite durable.

Here is the Montclair—the glamour car in this year's Mercury line. With its all-new styling and high speed it may become 1955's most popular car.

Thunderbird, Lincoln and Mercury

surely get to 30 mph in under 4.5 seconds and to 60 in under 12. And when you do tromp on the gas, you get an even surge of power transmitted through Lincoln's brand new Turbodrive automatic gearbox. This is a torque-converter unit similar to the ones in the Thunderbird and the new Merc that is coupled to a three-speed planetary gearbox and it gives acceleration so smooth you almost can't believe it. First-gear starts are possible without having to put the lever in low manually. All you do is floorboard the accelerator and the car takes off like a bomb in first and upshifts to second at full throttle. If you take off in a calmer fashion, you start in second which is not slow either. Any way you look at it, it's a real smoothie.

In spite of its big, powerful engine, the Lincoln should be an economical car. Last year's model, with 205 hp, was good for up to 18 mpg on the open road and gas mileage should be even better this year despite the horsepower increase. The reason is that the standard rear-axle ratio has been changed from 3.31 to 3.07. This is a wise move on Lincoln's part; it will make more difference in gas mileage than it will in acceleration. The 20 extra horses will see to that.

Standard equipment on all new Lincolns are tubeless tires—which seem to be taking the industry by storm—and twin exhausts. Prices for the Capri run from about $3,800 to $4,100 at the factory. If I had that kind of loot to spend on a car, I'd think twice before I bought anything else.

FROM the Lincoln I jumped right into a new Mercury Montclair—which may well turn out to be the sleeper of the year. It's hot as a pistol and roomier and better-looking even than the Capri (except for that you-know-what windshield). Back we went onto the giant soup-bowl track and here I got my first start. The car I was testing was so new it didn't even have all of its chromium on, which was fine with me except that it did look a bit naked. Anyway, there hadn't been too much care put into the body assembly because the car was only a test-grounds model. While I was circling the outside track at 85 mph with one of the company men, I suddenly heard a loud clattering noise as we came off one of the high banks. It was gone in a moment but I was sort of concerned.

"Don't worry," said the engineer, smiling. "It's only a hub cap."

I smiled back, wondering what was going to fall off next, but nothing did. In fact I kept smiling and the Merc took me up to an honest 103 mph just because I was so nice. Here again, there was no flat spot in acceleration and the car seemed so fast that I was anxious to run off timed tests.

So we did that next, before the handling and braking shakedown. And was I glad. Hang onto your hats, boys, because here is a rough idea of what a well-tuned 198-hp. Montclair can do: Zero to 30 mph took only 4.3 seconds, just two-tenths of a second slower than the Thunderbird! And that's not all. Zero to 60 time was 11.4 seconds, just one second behind the 'Bird with exactly the same engine (unofficially), and zero to 80 mph ran 20.7 seconds. These figures are all the more amazing when you consider that the Montclair is a full-sized sedan, weighing at least 350 lbs. more than the Thunderbird!

Still shaking my head in disbelief, I staggered back into the car after a brief rest period and headed for my old pal, the handling course. Here again the Merc performed like a slightly smaller Lincoln and I felt a definite advantage from the four inches less of wheelbase. The steering didn't have quite the positive feel of the Lincoln, but all the cars tested were early production-line models and the difference might just have been in individual cars. There was some body roll but nothing excessive and certainly not enough to affect control under ordinary driving conditions. If you want to go like a racing driver, you'll have to stiffen up the suspension of any American automobile.

Just as with the Lincoln, standard rear-axle ratios have been changed this year, from 3.54 to 3.15 on all Mercomatic cars.. And for the same reason as is true for the Lincoln, this will give a slight increase in gas mileage, in spite of the added horsepower. Speaking of Mercomatic, this year's version is better than ever and just about as smooth as Lincoln's new automatic transmission, with the same kick-down feature. Downshifts can be made into "low," by the way, at speeds over 50 mph, a useful factor for additional braking power.

Not that the Mercury needs any more braking power under normal use. I couldn't get it to show any fade at all and you shouldn't have any trouble either, unless you drive down from the top of Pike's Peak three or four times a day.

By now you may have gotten the impression that I like the Mercury. That would be an understatement. I love its looks, visibility, speed and handling. The available color schemes are startlingly brilliant, to say the least, but mostly in fine taste, and the interiors are cheerful and practical. I even like the new "frenched" headlight treatment that is featured on the entire Ford line. Every feature of these cars should be very popular, from their dual exhausts right on up. But don't let me be a salesman; go drive them yourself. You can walk away with a new Mercury for less than $3,000. It's something to think about.●

THUNDERBIRD SPECIFICATIONS
ENGINE: V-8, overhead valves; bore, 3.75 in.; stroke, 3.30 in.; total displacement, 292 cu. in.; developed hp, 193 with standard transmission, 198 with Fordomatic, both at 4,400 rpm; compression ratio, 8.1 to 1 conventional, 8.5 to 1 with Fordomatic; maximum torque, 286 ft./lbs. at 2,500 rpm; four-throat carburetor; ignition, 6 volts.
TRANSMISSION: three-speed manual or Fordomatic torque converter with three planetary gears; overdrive (opt.)
REAR AXLE RATIO: conventional. 3.73; overdrive, 3.92; Fordomatic, 3.31.
SUSPENSION: front, independent ball-joint coil spring system with two transverse control arms of unequal length; rear, semi-elliptic leaf springs with rubber-bushed shackles.
BRAKES: four-wheel hydraulic, internal expanding; 11-in.-diameter drums; power booster available.
DIMENSIONS: wheelbase, 102 in.; tread (front and rear), 56 in.; width, 70 in.; height, 52 in.; over-all length, 175 in.; turning circle, 36 ft.; weight, 3,150 lbs.; tires, tubeless, 6.70 x 15. (6.00 x 16 optional.)

LINCOLN CAPRI SPECIFICATIONS
ENGINE: V-8, overhead valves; bore, 3.94 in.; stroke, 3.50 in.; total displacement, 341 cu. in.; developed hp, 225 at 4,400 rpm; maximum torque, 332 ft./lbs. at 2,500 rpm; compression ratio, 8.5 to 1; four-throat carburetor; ignition, 6 volt.
TRANSMISSION: Turbo-drive torque converter with three planetary gears.
REAR AXLE RATIO: 3.31 standard; 3.07 with air-conditioning.
SUSPENSION: front, independent ball-joint coil-spring system; rear, semi-elliptic leaf springs.
BRAKES: four-wheel hydraulic, internal expanding; 12-in.-diameter drums; power booster available.
DIMENSIONS: wheelbase, 123 in.; front tread, 58.5 in.; rear tread, 60 in.; width, 77.6 in.; height, 62.7 in.; over-all length, 215.6 in.; turning circle, 45.7 ft.; weight, 4,300 lbs. (approx.); tires, tubeless, 8.00 x 15.

MERCURY MONTCLAIR SPECIFICATIONS
ENGINE: V-8, overhead valves; bore, 3.75 in.; stroke, 3.30 in.; total displacement, 292 cu. in.; developed hp, 198 at 4,400 rpm; maximum torque, 286 ft./lbs. at 2,500 rpm; compression ratio, 8.5 to 1; four-throat carburetor; ignition, 6 volt.
TRANSMISSION: Mercomatic torque converter with three planetary gears; standard three-speed manual and overdrive available.
REAR AXLE RATIO: conventional, 3.73; overdrive, 4.09; Mercomatic, 3.15.
SUSPENSION: front, independent ball-joint coil-spring system; rear, semi-elliptic leaf springs.
BRAKES: four-wheel hydraulic, internal expanding; 11-in.-diameter drums.
DIMENSIONS: wheelbase, 119 in.; front tread, 58 in.; rear tread, 59 in.; width, 76.4 in.; height, 58.6 in.; over-all length, 206.3 in.; turning circle, 42.8 ft.; weight, not available; tires, tubeless, 7.60 x 15.

ROAD TEST

*A great car gets even better
with performance improvements where they count the most*

Photos by Dean Moon

LINCOLN'S PERFORMANCE, particularly as winner of the big stock car class in the Mexican road race for the past three years, has put it securely into the championship class. When the same engineering design that formed the basis for these victories is further refined, as it has been in the 1955 Lincoln, the result calls for careful study.

As a matter of fact, it is in the vital power train itself that most of the major changes for 1955 have been made in Lincoln. Their influence was quickly noted during an 850-mile road test of the popular Capri hardtop. It is evident that Lincoln engineers chose to make slight sacrifices in maximum speed at the top end in favor of greater fuel economy and faster acceleration in the middle ranges.

Expressed in figures, the Lincoln gave up several miles per hour from 1954, but added this much and more in the 30- to 60-mph bracket. The choice obviously is a good one, since most drivers find passing ability useful, while maximum speed is rarely if ever attained in normal operation.

The contributing factors to these performance changes are in engine modifications, a new automatic transmission and a different rear axle ratio on all 1955 models, except those with air conditioning.

Lincoln's current engine, which reportedly is to be the basis for the forthcoming Continental, has an extra 20 horsepower over 1954, along with 32 more foot-pounds of torque. The displacement has been increased from 317 to 341 cubic inches, while the compression ratio has gone from 8-to-1 up to 8.5-to-1. Cylinders have been bored out to 3.94 inches, from 3.80 for 1954, but the 3.50 stroke is unchanged.

Other changes include: higher lift camshaft to decrease overlap and improve low-speed acceleration and smooth out the idle; increasing size of intake manifold passages by 10 per cent; repositioning choke plates and substituting 1⅛-inch secondary barrels in the four-barrel carburetor; going to the open wedge design in the combustion chamber.

NEW AUTOMATIC TRANSMISSION

The fire at GM's Hydra-Matic plant may or may not have been the opportunity Lincoln was waiting for to bring in its own automatic transmission. In any event, the new Turbo-Drive has the driving characteristics common to torque converters with planetary gearing. Take-off and easy acceleration are accomplished without the noticeable shift points. An extra feature is the low-gear start, even though the selector lever is in drive position when the throttle is floorboarded. Under these conditions, the testers noted the upshift coming at approximately 33 mph. Final shift point, with throttle wide open, is registered at 70 mph when the transmission goes into top gear.

When checked with a fuel-flow meter, the more powerful engine, in combina-

---

SPECIFICATIONS

Engine type	OHV V-8
Displacement	341 cubic inches
BHP	225 @ 4400 rpm
Compression ratio	8.5-TO-1
Bore	3.94 Stroke 3.50
Torque	332 ft.-lbs. @ 2500 rpm
Transmission	TURBO-DRIVE
Rear axle ratio	3.07
Wheelbase	123 inches
Dry weight	4510 lbs.
Turning circle	45.7 feet
Steering lock-to-lock	4.3 turns

PRICES

Car $3910	Power Steering $129
Transmission STANDARD	Power Brakes $43
Radio $131	Air Conditioning $621
Heater $121	

'55 LINCOLN CAPRI

MOTOR LIFE ROAD TEST

CAR TESTED: 1955 LINCOLN CAPRI

TEST CONDITIONS
- Altitude: 200 feet
- Temperature: 74 degrees
- Wind: 8 mph
- Gasoline: MOBILGAS PREMIUM

ACCELERATION AND TOP SPEED

MPH	0-30	0-45	0-60	30-50	40-60
Seconds	4.5	7.4	12.5	5.2	6

- Standing ¼ mile: 18.9 seconds
- Fastest one-way run: 107 mph
- Top speed avg. 4 runs: 105 mph

SPEEDOMETER CORRECTIONS

Car Speedometer	Actual Speeds
20	17
30	26
40	38
50	46
60	54
70	64
80	75
90	82
100	91

BRAKING DISTANCE

MPH	Stopping Distance
30	47 feet
45	89 feet
60	150 feet

FUEL CONSUMPTION

MPH	Average
30	21 mpg
45	20 mpg
60	16.5 mpg

REMARKS: ALL SPEEDS ACTUAL. PAVEMENT — DRY ASPHALT.

---

tion with the new transmission and a rear axle ratio of 3.07, showed a substantial increase in fuel economy of from two to four mpg at all speeds.

In recent years Lincoln has been unsurpassed among American cars for handling qualities at speed. Its competition record is evidence that with high speed must go the ability to stay on the road and get around corners. Since the Lincoln's suspension system is virtually unchanged, except for mounting of the rear shock absorbers at a more horizontal angle for a better ride and widening of the rear tread by two inches, these good points have been retained.

During the severe handling maneuvers, the road test crew noted an interesting fact. Pressure in the tubeless tires, checked before and immediately after, showed a drop of approximately three pounds. The explanation from engineers was that, under hard handling, some air may escape at the joint of rim and tire.

Although Lincoln's styling fundamentally was unchanged for '55, its lines were enhanced by extension of the rear fender line by some 10 inches. It remains the only car in its field without a wraparound windshield, but the practical value of this feature is highly questionable, anyway. The 1955 Lincoln is second to none in the "good looks" department. This, along with its performance and good ride and handling, makes it a good bet, both as a family vehicle and in open competition. •

Brake detonator is actuated by stop light which passes electric current to actuate charge fired onto road for marking.

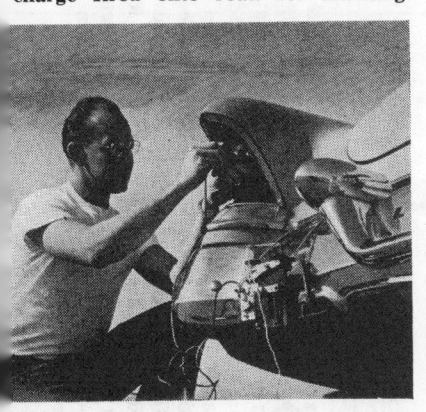

Tire pressure checks before and after severe braking tests showed loss of three pounds, a characteristic of tubeless type.

Outstanding feature of Lincoln is its clean, modern lines. Absence of chrome gimmicks should keep the resale high.

Lincoln's new headlight and tail-light bezels add to the long low look

LINCOLN

It's high-priced, but a unique combination of luxury car and road racer

Smallest of the big and powerful brute cars, the Lincoln should still go right on being the one that gets top money in tough competition like the Mexican Road Race. The reason is that Lincoln's suspension system, which it willed to the smaller Mercury and Ford, provides soft ride plus excellent cornering qualities that enable it to go into and come out of corners faster than either of its two principal sales competitors.

A new front grille, a new gold crest, and forward-raked headlight bezels with lines that make the car appear longer and lower are about the only exterior changes. Inside, the theme has been color, bright colors which do not clash and which harmonize with a new batch of exterior shades.

The biggest advancement in Lincoln for 1955 is the Turbo-Drive transmission; this marks the end of Hydramatic as far as the Ford Motor Company is concerned. Turbo-

Around the Dearborn high speed oval, the big Lincoln Capri steered hands off at 70 mph

■ LINCOLN

Drive is a torque converter hitched to a three-speed set of planetary gears—it's a larger edition of the automatics used in the Merc and the Ford, and it's built in Ford's new Livonia, Michigan, factory. There are five positions on the selector lever indicator (which is still on the steering column within easy view): beginning at the left is P (parking), then R (reverse), N (neutral), D (drive range), and L (low), so there's no need to go through a forward gear to get into reverse if you have need to rock the car in snow or mud to obtain traction.

With the selector lever in D, you get a second gear start *unless* you jam the throttle all the way down which gives you a low gear start. When you come to a sharp curve or a steep hill you can flip the lever to L and get a smooth and very quick down-shift for better control and engine braking. You can shift down thus at speed up to a shade over 60 mph, and the result is no longer accompanied by the small jolt that was often irritating. Under full throttle, automatic up-shift from low gear to second takes place at about 35 mph, then with throttle still wide open, up-shift to third is at about 75 mph, and all without touching the lever.

Lincoln engineers were interested in torque, so they raised piston displacement from 317 cubic inches to 341 by boring out the cylinders to 3.9 inches. The stroke stayed put, but compression ratio jumped from 8.0 to 8.5 to 1. Horsepower naturally went up 20 points to 225, but more important torque shot up from 305 foot pounds at 2300-2800 rpm to 332 foot-pounds at 2500 rpm aided by a new higher-lift camshaft. The result is that it's better to use premium grade gasoline, but you get a whale of a lot more go throughout the critical speed ranges of today's driving.

Also on the new Lincoln is the new conical-seat spark plug that features a wider heat range and larger air gap between the outside shell of the plug and the insulator. Not only is fouling decreased, but there's no gasket to worry about—the plug seats down better, too.

Lincoln engineers have provided an op-

84

SPECIFICATIONS
In inches unless otherwise stated

CHASSIS AND BODY
Wheelbase	123
Tread	58-½ front, 60 rear
Length overall	215-6/10
Width overall	77-6/10
Height overall	62-7/10
Ground clearance	6 (approximate minimum)
Turning circle diameter	45 feet 8 inches
Steering wheel lock-to-lock	3-½ turns with power
Tire size	8.00 x 15 (8.20 x 15 convertibles and air conditioned cars). All tubeless
Weight, shipping	4275 pounds
Brake lining area	220.06 square inches
Weight to brake area ratio	19.43 pounds per square inch
Weight to power ratio	19.0 pounds per BHP

ENGINE and contributing equipment
Cylinders, block, valves	8, 90-degree V, OHV
Bore and stroke	3.94 x 3.50
Displacement	341 cubic inches
Compression ratio	8.5:1
Brake Horsepower (maximum)	225 @ 4400 RPM
Torque	332 foot pounds @ 2500 RPM
Carburetor	Single, 4-barrel
Choke	Automatic
Fuel pump	Mechanical
Fuel recommended	Premium
Fuel tank capacity	20 gallons
Exhaust system	Dual with reverse-flow mufflers
Crankcase capacity	5 quarts (add 1 for filter)
Drive Shaft type	Exposed
Rear axle type	Hypoid
Rear axle ratio	3.07:1 (standard)
	3.31:1 (optional)
Transmission	Automatic only
Piston speed @ maximum RPM	2333.3 feet per minute
Electrical system	6 volts
Cooling system capacity	25-2/10 quarts (with dual-unit heater)

INTERIOR DIMENSIONS
	Sedans	Coupes	Convertibles
Hi· Hiproom	62-3/10 front,	62-3/10 front,	62-3/10 front,
	62-1/10 rear	59-8/10 rear	48-½ rear
H Headroom	35-½ front,	34-1/10 front,	35-1/10 front,
	34-7/10 rear	33-4/10 rear	35-3/10 rear
S Seat Height			
L Legroom	44-3/10 front,	44-3/10 front,	44-3/10 front,
	42-8/10 rear	38-2/10 rear	38-1/10 rear

PERFORMANCE Based on actual tests on all types of roads
Acceleration (standing start) to 30 MPH true speed: 4.8 seconds
to 45 MPH true speed: 8.0 seconds
to 60 MPH true speed: 12.7 seconds
Highway acceleration (with step-down) from 50 to 80 MPH : 12.2 seconds
Maximum speed 110 MPH plus
Weather during tests Sunshine, dry

VISION forward over hood (driver 5' 10") .20 ft. 6 in. forward of bumper

■ LINCOLN

tional built-in lubrication system (called Multi-Luber) to supply light grease, at the press of a dash panel button, to the chassis and suspension systems, steering, etc. An additional master cylinder under the hood supplies grease through nylon tubes which fit over the conventional grease fittings.

Though the rear tread has been increased to 60 inches, new location of the rear shock absorbers at a more nearly horizontal angle has increased rear end stability. Weight distribution has been good on Lincoln for several years, and after getting accustomed to the test Capri, I began plowing into corners at various speeds. The Lincoln leans, to be sure, but there's none of the tendency of the rear end to come around and meet you. The hairpin I used is out on the test track area inside the high speed oval. Normally entering this hairpin in a lower gear and throttling through at an even 25 mph is just about maximum for the average car. But in one of Henry's ball-joint jobs, this is like shooting fish in a barrel. In fact I made it through this corner, which is perfectly flat to add to the difficulty, at about 32 tire-squealing mph.

A quality car, contrary to popular opinion, need not be overly large. Little larger than many of the medium-priced range, and in fact smaller in wheelbase than a few of them, the Lincoln offers luxury combined with handling that seldom is found in the big breeds.

You can wind this baby up to considerably over 100 mph, but in city traffic the high compression engine is as docile as a spaniel. It's changed plenty this year, but lots of those changes like the redesigned combustion chamber that improves valve cooling and the elimination of the cross-over pipe from one side of the engine to the other are not noticed by the buyer. If you listen closely when you start a new Lincoln, you'll notice that even the tappets are silent.

Engine has four-barrel carburetor and dual exhaust system, gives 225 hp at 4400 rpm

Even the trunk's interior looks handsome. It's deep and wide, with an insulated lid

Brake pedal could be somewhat larger, but the power brakes grabbed less than in '54

HOW DO YOU DESCRIBE the feel of something? How it affects you, the impression it leaves with you? You can say something feels different, but it seems like an inadequate phrase. Yet that's the most striking thing about the '56 Lincoln. It *feels* entirely unlike any of the post-war Lincolns, from the moment you slip behind the wheel until you leave the car many miles later. Where the intervening models had an infectious informality, this car substitutes an aloof dignity that reminded me of my '48 Continental.

Well before the official announcement of the new Lincoln, Don MacDonald and I had the opportunity to drive an engineering car (the untrimmed black '56 model you see in the big photo) at the Ford Proving Ground in Dearborn, Mich. My impressions of the car (in bold type) are embellished with comments and further explanatory copy by Don.

One of the 1st things I noted as we rolled out of the garage (aside from the feel) was the extreme quiet and lack of road rumble.

Lincoln considers svelte smoothness, with its attendant silence, of extreme importance in this 1st serious attempt to dislodge Cadillac from its corner on the fine-car market. Lift the hood and you will see that the engine is packed like glassware in a compartment wholly topped with a new-type Fiberglas and woven paper insulation. Much attention has been paid to isolation of body from frame. This, coupled with softer springs and shocks, will cushion you in the best boulevard tradition.

Instrument legibility is slightly better than previously, mostly because of the contrasting background. I hope the '55 white on white (of one of our test cars) won't be repeated on any of the new models.

first time around '56

the '56

by Walt Woron and Don MacDonald

The basic instrument design, with its mammoth horizontal-sweeping speedometer dwarfing other gauges aligned alongside, has been a feature since 1952. It is functional, and well adapted to integral placement of accessory controls. Unlike other Ford Divisions who are also stressing safety for 1956, Lincoln is settling for a non-padded, vinyl-covered panel. A padded panel can only make a very limited contribution to an overall safety program, but however limited, we consider it essential. We predict thru public pressure that it will be available as an option on '56 Lincolns produced later in the model year. Meanwhile, the non-glare vinyl cover is a step in the right direction, even tho the 1st 100 cars produced (as in all our photos except those of the convertible) lack this feature.

The rest of the new Lincoln interiors emphasize luxury with no sacrifice in safety. The insides of the top-line Premiere are based on bolsters of genuine leather color-keyed to quite rich-looking and tweedish cloth seating areas. All-leather is a plugged option in the hardtop and convertible, whereas the now "standard" Capri closed models come upholstered throughout in fabric.

Power windows and front seat are standard on the Premiere, optional on Capri. We particularly like placement of the window controls to the left of the driver on the flange of the instrument panel underneath the new wrap-around area of the windshield. Safety belts for all passengers are another featured extra, tying in with current safety statistics showing that motorists stand a better chance when held to their seats in a serious accident. The occasional freak accident where a person gets thrown from his car into a rosebush and stands up smiling does not alter the twice-as-good chance of survival inside in a crash.

CONTINUED

LINCOLNS

Definite family resemblance to the '55 model, but younger, cleaner-cut

Lower and immensely longer (from rear especially), '56 is impressive

PHOTOS BY JOE FARKAS

Pole position in this year's introduction race goes to one of the few entirely new '56 models, a car that looks, acts, and even feels different

Futura's shadowed eyelid is used, not noticeably more modest, with double parking lights (future headlights?)

With the largest unbroken areas of any new car but one more expensive one, Lincoln has great look of space

(Left) Roaring down the trough of test track at Dearborn, stripped engineering version was sure-footed

First serious use of a rear-mounted, production car "grille," this one teams with big exhausts for harmonious rear end

87

The '56 Lincolns

continued

Very big, very impressive, Lincoln Premiere sedan will attract whole host of new buyers. Strange triangular window at rear should help air conditioning start work

Lavish roominess of convertible is not evident immediately because of skillful proportioning, fine detailing of interior. Note safety wheel, power window control off door

The handgrips on the inset wheel (also part of the crash and safety program) were in the right place for me, and I felt that they would stay comfortable on a long trip.

Lincoln's new steering wheel is more than just decorative and comfortable to hold onto all day. You will note that 4 "safety-flex" spokes raise the rim a good 3½ inches from the hub. We saw movies where a 150-pound weight (a compliment to the average driver) was swung on a pendulum against the new wheel. Rather than impaling itself on the hub (a prominent cause of accidental death in other cars), the weight came to rest with at least an inch of spoke play left. The new design effectively prevents unintentional hara-kiri via a sharpened steering column.

Before we got to the acceleration strip from the garage I had to use the brake several times. The width of the pedal was advantageous for either left- or right-foot braking. When I asked the Lincoln chassis engineer accompanying us for the reason behind the 2 lever arms straddling the column, he explained that their experience indicated foot pressure on the pedal was less apt to bend it when it had 2 levers.

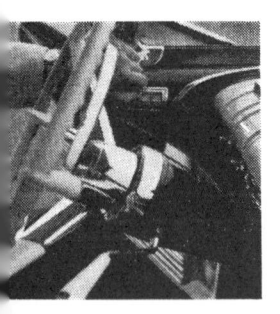

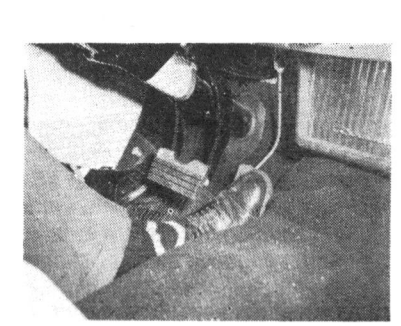

Portrait of a safe car to ride in: top visibility to all sides, instruments near eye level, recessed wheel, safety belts. New stay-shut doorlatch would pay off if bad crashes should occur

Control of all major windows is handier than in most cars, since driver uses buttons near wheel. Rugged power brake pedal is sturdiest we know of, demonstrates industry's new interest in safety

88

Heat and vent controls (at left) are positive and even fun to work, with notification of interior climate appearing in the twin dials. Toggle switches carry out the airplane feel. Excellent design has extended to radio grille and both pedals, has missed exasperating glove compartment

The new cowl-suspended, double-armed brake pedal and cowl-mounted accelerator pedal follow current trends. On the accelerator in particular, it allows a much-simplified linkage to provide more positive response. Today's carburetors are complex enough without routing the control circuit under the floorboards and up. In both cases, it simplifies service because all units are readily accessible under the hood.

There's a definite improvement in acceleration, altho not noticeable by the normal "seat of the pants" method. We had a match race between the '55 and '56 (more for photo comparison than technical check) and noted an immediate difference from scratch to the end of the strip (around 80 mph). Flicking the shift lever down to take advantage of the low-gear start feature without the disadvantage of waiting for the downshift, we were able to get the following figures: 0 to 30 mph in around 4½ seconds (about the same as last year), and to 60 mph in around 11½ seconds (one second better than the '55). The reason for the "about" is that we did not use a 5th wheel, were using average speedometer error instead, and had only one stopwatch in place of the bank of watches we use in our more accurate and thorough road tests (to come later).

At cruising speeds, acceleration is considerably improved, the car taking less than 4 seconds to get to 50 from 30 (nearly 1½ seconds better) and under 12 seconds to get to 80 from 50 (2 seconds faster than in '55).

The increased horsepower (up 60 hp from '55 to 285) is largely responsible for this, since the transmission and rear-axle ratios are the same.

Since 1952 when Lincoln produced the first ohv V8 in the Ford family, the parent company and its divisions have astonished us each year by coming up with an engine so new that hardly any parts are interchangeable. (Proof of this pudding is outlined in Al Kidd's article on engine modernization, starting on page 24.) This year's whopping increase in power (percentage-wise, torque increase is even greater with a jump from 342 to 401 pounds-feet) required a new block, crankcase, induction system, and 9 to 1 compression ratio. In other words, you can be all-new and still be a relative. As for other changes, a 4.0-inch bore and a 3.66-inch

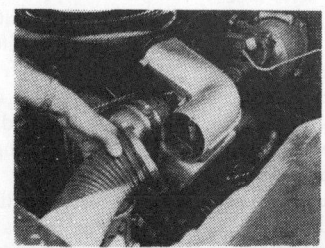

Power steering, now standard on every Lincoln, is integral with steering column except for pump. Compatible air is assured at carburetor with simple heat valve in air supply, promising both faster warmup on frosty mornings and freedom from overheating in the summer. Engine's gimmicks are all simple

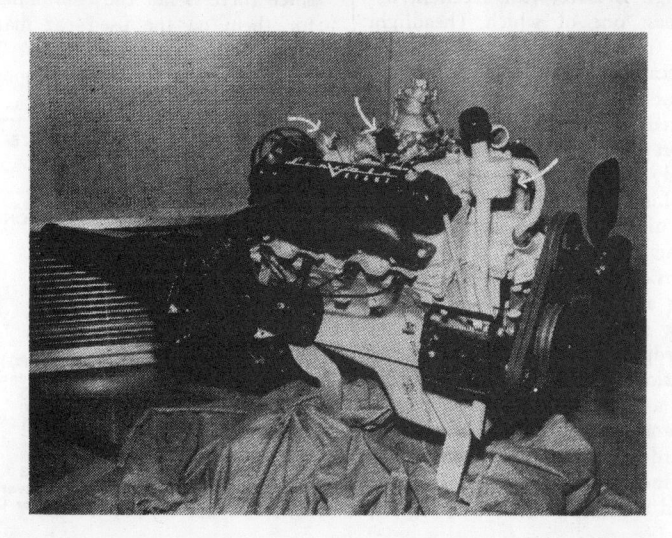

Tho it follows general design of overhead-valve Lincoln introduced in 1952, the '56 V8 is larger (368 cubic inches), displays big torque increase

89

The '56 Lincolns

stroke give 368 cubic inches of displacement vs. last year's 341. Obviously, this required redesign for structural strength and increased cooling. The new crankshaft has such good inherent balance that the usual 2 center counterweights were eliminated. Rpm for peak horsepower has been increased to 4600 mainly due to a higher-lift cam, beefed-up valve springs, and better breathing thru a completely redesigned induction system. Larger exhaust valves are serviced by a more efficient dual exhaust system.

Little details that will eventually save you repair money include a rotary oil pump that, unlike the old gear type, will put out peak pressure even when badly worn; an all-new crankshaft ventilation system which is forced-draft proof against acid accumulation; and a cam whose shaft is much thinner than the least dimension of the lobes, allowing close control of heat-treating processes and therefore better cam-to-tappet compatibility.

The '56 Lincoln handles unlike the '55 model. We felt (May '55 MT) that the previous car was still near the top in roadability without any sacrifice in riding comfort. In softening up the ride for this year, Lincoln engineers have not been able to keep the same roadability it had before. Don't get me wrong; it's still plenty good. Now, tho, instead of drifting evenly thru a turn taken fast, it breaks loose quicker at the rear end and you have to be on your toes. One engineer argued that the '56 is a car you have to force into the turns, whereas the '55 walked right into them with no coaxing. It may be true, but after driving a '55 around the same twisting asphalt course just prior to taking the '56 around, I don't believe it. Driving normally, you won't notice much difference between the 2 cars.

There seemed to be slightly more body lean when taking corners, but it is not uncomfortable. At high speeds the car is stable and sure-footed.

Lincoln engineers like to think that all of their product's famed roadability has not disappeared down the boulevard, and will challenge you at the drop of a hat (you drive the '55) from any point A to any point B. Even tho Walt Woron's initial impression was that the new car sacrificed for ride some of the qualities that have made previous models consistent Mexican Road Race winners, he admits that he could be wrong and will reserve judgment until our full-scale road test.

The chassis under the new body is also entirely new, but quite orthodox. Wheelbase has been stretched 3 inches from last year's 123; overall length a little more than 7 inches, mostly in the back-end overhang. Road clearance is an immaterial ½-inch less (still greater than current Cadillacs) but overall height has been dropped a good 2½ inches (to 60.2) without appreciable sacrifice of headroom. The new car weighs roughly 200 pounds more than the old.

Suspension, basically the same as last year except for softer rear springs, is bound to feel different when supporting these new dimensions. As we said before, Lincoln wanted a boulevard ride, and juggled details to obtain it. The new ride can be criticized only in the light of the superlative roadability of the old, and we are not too sure as yet that any criticism is valid.

Lincoln's power steering, which is standard on all models, is an entirely new design. In fact, their 1st with it has caused

"Just think of it! No more fuel bills!"

some embarrassment to GM's Saginaw Division, who is the supplier. It is somewhat similar to Chrysler's in that all units except the pump are housed in the steering column. Retained, tho, is feel-of-the-road with roughly 3½ turns lock to lock.

Some purists will say that style-wise the new Lincoln has lost its individuality. The majority will say that it is a very good-looking car, perhaps without knowing why. The reason is simply that Lincoln stylists aimed at producing a low-slung, massively wide car that was physically big but looked even bigger. They took a bead on what buyers of luxury automobiles seem to want today, and created same.

The design is alive with excellent details, at least one of which (headlight treatment) is straight from the Futura. Perhaps because wrap-around windshields have become so commonplace, Walt Woron neglected to mention increased visibility (more glass area than any other car in its field) from this overdue Lincoln feature in his driving impressions. Use of chrome is nicely restrained to functional-seeming bits that tend to increase whatever dimension is their plane. The grilles, both front and rear, are massively neat.

The now standard Capri is available in 2-door hardtop and 4-door sedan form. The luxurious Premiere line includes these, plus a convertible. Turbo-Drive automatic transmission and power steering are standard on both. Every conceivable accessory, including air-conditioning, can be installed at the factory for a price.

Lincoln Motor Trials

CONTINUED FROM PAGE 23

almost three decades of experience with production of the V-8 engine. The Leland Lincoln became Ford's first V-8, and Ford has been the world's biggest producer of this type of engine, has had years of experience in acquiring, and developing know-how—a pleasant position to be in as the automotive world awakens to the superiority of the V layout.

As far as I know, not even its manufacturer calls attention to the fact that Lincoln's engine is the biggest being fitted to a passenger car today, anywhere in the world—the reason being, I suppose, an understandable desire to avoid creating the impression of a gas-eating gargantua in the economy-conscious public mind. However, economy-wise, a big engine lightly stressed is equal to or better than a small engine pushed to its limit.

The Lincoln engine reeks reliability; it's a simple, un-gadgety design that has been refined to the ultimate degree over the years. Outstanding features are its forged crankshaft (not cast, as in the Mercury), excellent crankcase ventilating system, fore and aft vibration dampers (the flexible flywheel doubles in this capacity), and hydraulic valve lifters. There's nothing more annoying than tappets that don't tap, and we've encountered them in more than one hydraulic-tappet engine. We were pleased to find that, in spite of deliberate high over-revving, the Lincoln's tappets did their job properly and in silence. They operate at zero clearance, regardless of valve condition or engine temperature, and require no adjustment. Like the rest of the car, they're made to serve silently and faithfully.

TREND TRIALS NO.: The Lincoln "Sedan Sports" falls in the $2501-2700 price bracket and its Trend Trials No.—index of operating expense and depreciation—works out to 40.48, a reasonable figure for a quality vehicle. Large, quality cars do have a higher depreciation rate since there is not the consuming demand for them on the used car market that exists for economy machines. The T T No. is also boosted by the proportionately high cost of replacement parts.

TABLE OF PERFORMANCE
DYNAMOMETER TEST
1200 rpm (full load) 30 mph		45 road hp
2000 rpm (full load) 49 mph		71 road hp
3100 rpm (full load) 75 mph	(max.)	95 road hp

ACCELERATION TRIALS (SECONDS)*
Standing start ¼-mile	:19.90
0-30 mph (no gear change)	:05.49
0-60 mph through gears	:15.58
10-60 mph in high gear	:23.42
30-60 mph in high gear	:12.83

TOP SPEED (MPH)*
Fastest one-way run	100.67
Average of four runs	97.08

FUEL CONSUMPTION (MPG)
	Conventional	Overdrive
At a steady 30 mph	22.13	26.53
At a steady 45 mph	20.00	24.98
At a steady 60 mph	15.61	17.05
Through light traffic	18.33	
Through medium traffic	15.03	
Through heavy traffic	11.33	

*Without use of overdrive. High gearing permits identical top speed in conventional or OD.

Lincoln Motor Trials

BRAKE CHECK
Stopping distance at 30 mph	44 ft. 11 ins.
Stopping distance at 45 mph	101 ft. 3 ins.
Stopping distance at 60 mph	238 ft. 0 ins.

GENERAL SPECIFICATIONS

ENGINE
Type	L-head V-8
Bore and Stroke	3½ x 4⅜
Stroke/Bore Ratio	1.25:1
Cubic Inch Displacement	336.7
Maximum Bhp	154 @ 3600
Bhp/Cu. In.	.457
Maximum Torque	265 ft. lbs. @ 2000 rpm
Compression Ratio	7:1

DRIVE SYSTEM
Transmission—Conventional three-speed. Ratios: First—2.526:1, Second—1.518:1, Third—1.0:1, Reverse—3.158:1
Optional Overdrive—.722:1
Hydra-Matic Ratios: First—3.8195:1, Second—2.6341:1, Third—1.4500:1, Fourth—1:1, Reverse—4.3045:1
Rear Axle—Semi-floating, Hypoid, Hotchkiss drive. Ratios available: 3.31:1 (Plains), 3.91:1 (Standard), 4.27:1 (Station Wagon).

DIMENSIONS
Wheelbase	121 ins.
Overall Length	214 ins.
Overall Height (loaded)	63.6 ins.
Overall Width	76.7 ins.
Tread	Front—58.5 ins., Rear 60.0 ins.
Turns, Lock to Lock	5½
Weight (Test Car)	4375 lbs.
Weight/Bhp Ratio	28.4:1
Weight/Road Hp Ratio	46.0:1
Weight Distribution (Front to Rear)	55.4/44.6

'54 Lincoln Road Test

CONTINUED FROM PAGE 65

GENERAL SPECIFICATIONS

ENGINE: Ohv V8. Bore 3.80. Stroke 3.50. Stroke/bore ratio 0.92. Compression ratio 8.0 to 1. Displacement 317 cu. in. Advertised bhp 205 @ 4200 rpm. Bhp per cu. in. 0.65. Piston travel @ max. bhp 2450 ft. per min. Max bmep 97.5 psi. Max torque 305 @ 2300-3000 rpm.

DRIVE SYSTEM: AUTOMATIC is Dual-Range Hydra-Matic, fluid coupling with gears. Ratios: 1st 3.81, 2nd 2.63, 3rd 1.45, 4th 1.0, reverse 4.30.

REAR AXLE RATIO: Hydra-Matic 3.31 standard.

DIMENSIONS: Wheelbase 123 in. Tread 58.5 front, 58.5 rear. Wheelbase/tread ratio 2.1. Overall width 77.44 in. Overall length 214.8 in. Overall height 62.7 in. Turning diameter 45.7 ft. Turns lock to lock 4.25. Test car weight 4480 lbs. Weight/bhp ratio 21.8. Weight distribution 57% front, 43% rear. Tire size 8.00 x 15.

PRICES

(Including suggested retail price at main factory, and delivery and handling charges, but not freight)
COSMOPOLITAN: four-door sedan $3537, hardtop coupe $3640. CAPRI: four-door sedan $3726, hardtop coupe $3884, convertible $4045.

ACCESSORIES: Hydra-Matic standard. Power steering $145. Power brakes $40. Radio $122. Heater $113. Air conditioning $647. White sidewall tires (exchange) $37. Power windows $65. Power windows with four-way power seat $165.

ESTIMATED COST PER MILE

The estimated Cost per Mile is given for comparative purposes and is intended only as a guide to the cost of operating a particular car. To provide this comparison, Motor Trend has taken into consideration all factors affecting operating costs and ownership costs. All figures given are the average costs nationally. For complete explanation of our method of determining the various figures, see "What It Really Costs to Operate Your Car" in July '54 MT)

OPERATING COSTS:
Gasoline	$220.00
Oil	14.85
Lubrication	19.50
Oil filter	6.24
Wheel alignment and balancing	13.20
Brake relining and adjustment	16.57
Major tune-up	14.00
WHAT IT COSTS PER MILE TO RUN	3.0¢

OWNERSHIP COSTS:
Sales tax and license fees	$105.30
Insurance	173.40
Estimated depreciation	590.00
WHAT IT COSTS PER MILE TO OWN	8.7¢
TOTAL PER MILE COSTS IF YOU PAY CASH	11.7¢
Finance charges	175.00
TOTAL PER MILE COSTS IF YOU FINANCE	13.5¢

PARTS AND LABOR COSTS

(These are prices for parts and labor required in various repairs and replacements. Your car may require all of them in a short time, or it may require none. However, a comparison of prices for these sample operations in various makes is often of interest to prospective owners. First price is for parts, second for labor.)

Distributor $11.95, $3.50; battery $22.95; fuel pump $17.48, $1.40; valve grind $2.00, $33.60; one front fender $57.05, $23.45; bumper $61.90, $5.60; two tires $61.72; Total parts $235.05, labor $67.55.

CONTINUED FROM PAGE 70

All are perfectionists, without the neurotic overtones that sometimes accompany this attitude; to them, perfectionism means something very simple: if the job isn't perfect, they just keep at it until it is. That's why they had a better chance to win than anyone else.

So that's how they got ready. Now we'll see how they do on the 23rd.

Pete

MEMO from Pete Molson
11/23/54

To: Walt Woron

Looks like we guessed right, at least in part. It's true that Ray Crawford wasn't officially a part of the factory team, but it's just as true that Stroppe and Smith prepared his Lincoln in the same way they did the others. With a time of 20 hours, 40 minutes, 19 seconds, he was a minute and 48 seconds ahead of Walt Faulkner, who in turn led Keith Andrews' Cadillac by just a minute and seven seconds.

Ferrari took the big sports cars (Maglioli, Phil Hill, and Chinetti in that order); Porsches driven by Hermann and Juhan led Chirno's Osca in the small sports; and Dodges swept the small stocks (drivers were Tommy Drisdale, C. D. Evans, and Ray Elliot). And that does it for this year.

—Pete Molson

CONTINUED FROM PAGE 43

The premier racing event of the North American continent is the tortuous Mexican Road Race, which starts near the border of Guatemala and finishes at Juarez, near the Texas border. The 1952 event in the stock car division was won by Chuck Stevenson driving a 1953 Lincoln. His average speed for the 1934-mile, danger-filled course was 90.96 miles per hour. Three other Lincolns captured second, third, and fourth positions.

LINCOLN ENGINE IMPROVEMENTS
	1952	1953
LARGER INTAKE MANIFOLD RUNNERS	1.56 SQ.IN.	1.80 SQ.IN.
LARGER INTAKE VALVE DIAMETER	1.74 IN.	1.98 IN.
HIGHER INTAKE VALVE LIFT	.3375 IN.	.3545 IN.
HIGHER COMPRESSION RATIO	7.5 TO 1	8.0 TO 1
INCREASED IDLE VACUUM	16 IN. HG.	17.5 IN. HG.

A new feature for the 1953 Lincoln is a second door-stop position. Door can now be stopped half-open.

Exclusive Report

Rebirth of the Continental

By G. M. LIGHTOWLER

I'VE just driven America's most exciting car—the Continental Mark II. And it's all automobile, every inch of its $10,000 218.4-inch overall length.

Edsel Ford would be proud of his son Bill, the youngest of the Ford brothers and the man most responsible for carrying this new Continental project through to its fabulous culmination. For the impact of this auto will be felt in every country that makes automobiles. It will be felt also by every person who considers his car more than were transportation. Its simple distinctiveness of line and air of supreme elegance will be sought after by all who have an appreciation of quality and regality.

Apart from the advancements made in the field of power equipment, automatic transmissions and the torsion bar suspension system, the introduction of the Continental Mark II is about the biggest thing that has happened in Detroit since the evolution of the Model A. It is fitting that both these achievements should be accomplished by the same firm, for family relationship is inherent in the latest Continental.

Before describing the thrilling experience of testing this latest model let us consider the history of the Continental line, for the Mark II reflects much of what has gone before, despite the fact that this is an ultra modern automobile.

Edsel Ford had a great appreciation of the arts, a fine understanding of human beings and displayed great mental flexibility and powers of visualization. Although he was a good business man, he was not overwhelmed by the industrial age's clamor for undivided attention to production and finance; he loved good music, good plays, fine sculpture and classic paintings. However, he was no antiquarian in his ideas and he foresaw the possibility of blending the best of tradition with the best of modernism. He did not aspire to the philosophy that

Chrome trim is conspicuously absent. Continental's lines lack sporty look.

The most eagerly awaited new model in Detroit history finally has been released to the car-buying public. CAR LIFE presents this first driving impression of the fabulous Continental Mark II

The long, elegant hood, plus the distinctive Continental spare, make Ford's prestige car something to wish for. In case you're interested, it's a $10,000 creation, truly America's most wanted car, fulfilling the fondest dreams of Edsel Ford.

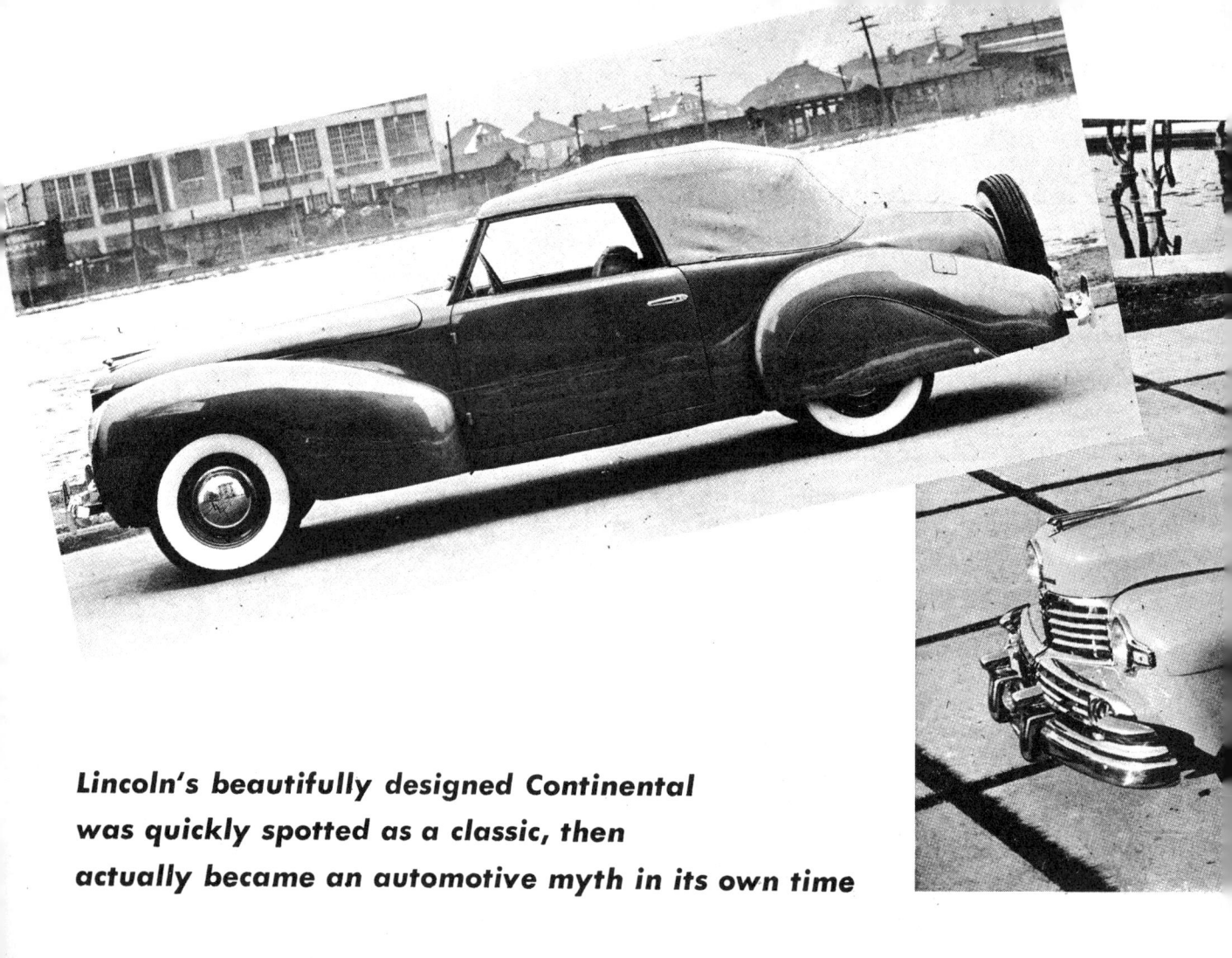

Lincoln's beautifully designed Continental was quickly spotted as a classic, then actually became an automotive myth in its own time

just because a thing had to be produced at a cheap rate and that it had to make a profit on sale it had to appear cheap. Although his ideas did not influence cars produced for the general public, they did dictate the requirements of his own personal cars.

He returned home from Europe with a burning desire to build a truly distinctive American car embodying some of the better features he found in foreign automobile design. His idea was to utilize the various components of the current Lincoln-Zephyr, and he personally supervised the production of a prototype along these lines. His original intention was to construct but two or three of these cars for himself and his elder sons, without serious thought of going into even limited production with a vehicle that required so much time-consuming attention.

After his vacation in Florida, it was decided to go into limited production of the Lincoln Continental. Altogether more than 5,000 models were sold to highly discriminating motorists throughout the world.

Before the public could fully appreciate the significance of the Lincoln Continental the world was plunged into the second World War and the automotive industry was more concerned with production of tanks and war equipment than fine automobiles. Immediately after the war the manufacturer was forced to give his full attention to the more mundane problems of producing a selling car for the public, and refinements had to take a back seat so that high-volume technique could be accommodated. In 1948 it was decided to suspend production of the Continental.

It soon became evident to the Ford Company (and it seems strange that it wasn't previously appreciated) that there was a growing demand for the return of the Continental. Inquiries that flowed into the company not only came from previous Continental owners but also from the many persons who had hopes they would one day be in a position to own a car of such distinction.

Public opinion eventually had its reward and in 1953 the Ford Motor Company decided to give serious consideration to the rebirth of the Continental, with William Clay Ford heading the organization that was to be entrusted with this operation, for not only was Bill on the Florida trip when his father decided to put the Continental on the market, but he also mirrored his father's personality more closely than did his brothers.

The staff of the Continental Division was selected with particular care, for each member had to have a sincerity of purpose. This sincerity was most evident in the three persons who formed the basis for the project: Bill Ford, who dearly wanted to carry on the job his father had started; Harley F. Copp, the production engineer who believed in an extremely high standard of automotive engineering, and John M. Reinhart, the chief stylist who firmly adhered to Edsel's philosophy of the blending of the best of tradition with the best of the 20th century.

The result of their joint efforts is astonishing, and if these three never produce another item of value they will have served the motoring public of the world far more faithfully and successfully than a great many of their predecessors and successors.

It must be remembered that this is a high priced car, higher priced than any other automobile manufactured in the United States for straight sale. It is also going to be produced in small quantities, probably only at a rate of five a week. Although it is not going to be a car that anyone can buy, either because of cost or on the grounds of availability, it is going to be the car that most motorists would want to own.

Style-wise the 1955 Continental Mark II is going to stop everybody on the street unless he is completely blind or walks around contemplating nothing but

The evolution of a super-car is shown here. Left is the 1939 Continental, center is the '48 coupe, and below is the new version of the Continental.

The Mark II, a name the new Continental will bear until a new body design is introduced in about four years, incorporates all the modern equipment usually considered as extras. It combines looks, with a high quota of luxury and liveliness.

Shadows show the extended width of the Continental chassis.

The left tail light assembly swings open to reveal gas filler cap.

the cracks in the pavement. A great deal of thought and hard work have gone into styling of the Continental, for although the new car had to bear a resemblance to the 1939 and 1948 models it still had to stand up against modern conception. It now shows its parentage and at the same time exudes a dynamic personality of its own.

The clean lines of the car will be appreciated immediately as will the absence of ornamentation and superfluous chrome. The lines of the car are both pleasing and functional.

The shape of the car is reminiscent of previous Continentals, and features like the Continental tire, copied by practically every other car manufacturer in the United States, has been retained in modified form. The rear deck that was such a significant feature of the 1939 and 1948 models, plus the long, elegant hood, have been repeated in modern guise. Although the car looks powerful and speedy, it does not have a "sporty" air. It has the high-performance-with-dignity-look. To avoid the "sporty" atmosphere, wire wheels are not used. Instead, it has the conventional wheel

(Continued on Page 98)

Four-barrel carburetor works normally on two barrels but other two activate on demand.

The Tape Line

Here are the specifications of the Continental Mark II:

Engine—Overhead valve V-8; displacement, 368 cu. in.; bore and stroke, 4.00 in. x 3.66 in.; compression ratio, 9.0 x 1.

Transmission—Torque converter; gear ratios, first, 2.4; second, 1.467; third, 1.0; reverse, 2.0.

Power Steering—Recirculating ball; gear ratio, 19.0 x 1; overall ratio, 22.1 x 1.

Wheels—Disc drop—center rim; rim width, 6 inches; tires, 8:00 x 15 tubeless, 4-ply.

Brakes—Hydraulic, internal expanding, duo-serve, single anchor. Effective braking area, 207.69 square inch.

Exterior Dimensions—Wheelbase, 126 inches; overall length, 218.4 inches; overall height, 56 inches; overall width (maximum), 77.5 inches; tread, front, 58.57 inches; rear, 60.0 inches.

Interior Dimensions—Head room, front, 35.4 inches; rear, 34.3 inches; leg room, front, 44.8 inches; rear, 43.9 inches; shoulder room, front, 59.4 inches; rear 58.4 inches; hip room, front, 61.8 inches; rear, 63.8 inches.

Price—Around $10,000.

Four-dial instrument panel presents a clear picture as to what's going on.

Here's the most attractive wraparound on the American auto market today.

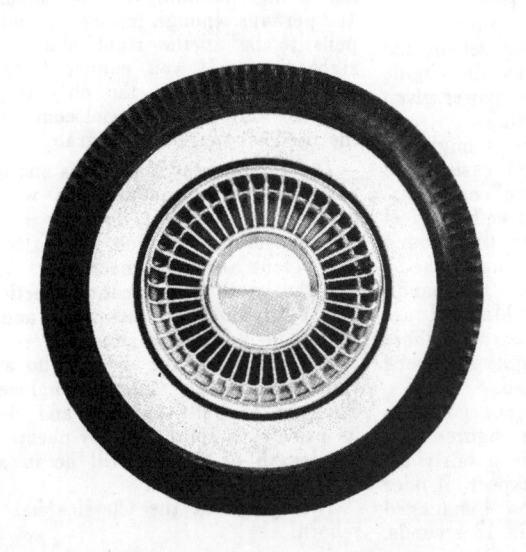

97

The New Continental

(Continued from Page 96)

with a specially designed hub with vanes. The wheels are indicative not only of the care exercised in design, but also the attention given to construction.

The design of the wheels and the hubs occupied much time before it was discovered that pressing them from a single metal sheet would not give the desired result. The vanes are all made individually and are bolted onto the center of the hub and the outside hub rim. The effect is excellent. The wheels themselves are carefully balanced both statically and dynamically with weights fitted both on the outside of the wheel rim and on the inside of the rim.

The front fender ventilation louvres exemplify how the idea of only functional design was followed.

An early design of the car provided ventilation louvres fitted to the bottom rear of the front fender. These were deemed necessary to dissipate some of the heat from the engine. The design effect of these louvres was most striking. However, in operation it was found that these cuts did not help carry heat to the outside, so they were eliminated although they looked nice. If the louvres were to be phoney, they had no place on the Continental Mark II.

The wraparound windshield fits into the general scheme without causing any disfigurement of grace. It seems likely this will be the first feature to be copied by other manufacturers.

The luggage trunk is not large, as the manufacturers felt this car wouldn't be taken on long trips. With this we do not agree, for we are certain that when one owns a 1955 Continental he is going to take it everywhere he possibly can. The space in the luggage trunk is curtailed further by the positioning of the spare tire, which blocks the entrance. The obvious place for the spare in a car of this nature is on a tray fitted beneath the luggage trunk and with its own hinged door.

The interior of the car is as luxurious and elegant as any American automobile can be in this day and age. There are many design innovations and the make up of the door itself is the most conspicuous. The doors are thicker at the base than at top. A shelf that runs the full length acts as an arm rest, fitting for the ash trays and power window lift and seat controls, and also as a handle to pull the door closed. The ventilation system is carried through a channel in the base of the door and sends warm air into the rear compartment through a vent in rear of the door.

The upholstery is in two-tone high grade leather and well-wearing felt of matching color. Where chrome is used, and there is quite an amount on the inside of the car, it is of extremely high quality.

The instrument panel conforms to the theme of "modern formal" and consists of four dials set closely together in front of the driver. On the left is a composite dial of four smaller guages for fuel tank capacity, oil pressure, temperature and amperage. To the right of the composite dial is the speedometer, graduated up to 140 mph. and at its right is a clock with a second sweep hand. Over to the right of the group is the calibrated tachometer.

The aircraft-type design of the heater and air controls is an interesting feature, with the throttle styled levers being operated from a panel above the transmission housing and in easy reach of the driver.

Like all Ford Company cars for 1956, the Continental embodies such safety features as the safety lock, the safety steering wheel, padded tray above the control panel, safety sun visors and the safety mirror. Safety belts will be optional equipment.

On all Continental Mark IIs full power equipment, automatic transmission, power steering and power brakes, station-selecting radio and heater will be standard equipment. Air conditioning will be optional and when fitted will be built into the car so that no offensive plastic or metal channels will be visible. All ducts will be behind the panel with four outlets built into the roof covering.

Sitting in the car one is immediately impressed by the excellent driving position. Although the hood is long, both fenders are clearly visible and one loses the feeling of expanse. All around visibility is excellent and although there is slight distortion from the wraparound, it is nothing like that experienced on contemporary cars. The swept area of the wraparound is immense, and the dirt left on the lower reaches (an area normally not reached by the wipers) does not cause any great trouble.

The starting procedure is conventional, with the transmission lever set in the "neutral" position. But once the engine fires convention ceases. The power given off by the motor is thrilling.

Take-off is not as fierce as might be expected, although one can easily spin the wheels in either "drive" or "low."

For those not acquainted with the feel of a car that sits close to the ground the Continental Mark II may take a little while to get used to. The handling characteristics of the Mark II are strictly of the Lincoln variety and there is no feeling of danger despite the speed at which the car is handled.

The manufacturers are not going to divulge any performance figures but after our tests we can give a fairly accurate idea as to what to expect. Under not too favorable conditions, I managed to zip the car to 60 mph. in 13 seconds.

It proved to be capable of 125 mph. plus! The engine has a capacity of 368 cubic inches and employs a 9:1 compression ratio. From our experience at the wheel, and after manipulations of the slide-rule and judicious questioning, we believe that the engine develops slightly under 290 bhp. at just over 4,500 rpm. We believe that the torque output is close to 400 foot pounds at near 2,800 rpm. The Continental Division will not definitely agree to these figures, but it will not deny that we are close.

There has been much speculation on the manner in which the Continental Mark II will be marketed. Most of the opinions expressed on this matter have been completely off-beat and based on anything but fact. The sale of Continentals will neither be controlled nor directed, other than by the simple restriction of availability.

The Continental Division has asked all Lincoln-Mercury dealers to indicate whether or not they would like to handle the new car; taking into consideration that they will have to equip their service departments with special tools to maintain the cars, and that they will not be handling anything like the large quantities of Lincolns and Mercurys. This means that to begin with a dealer will give considerable thought as to whether he wants to handle the Mark II, for it is most unlikely that more than three or four of the cars will pass through his hands during the first year of distribution.

A dealer, once granted the privilege of selling the Continental, and realizing the exclusiveness of his product, will naturally sell the cars to the most influential customers in his area. Already enough deposits have been paid to potential dealers to guarantee the sale of a very large portion of the first and second year's production of the Mark II. Among those who have virtually guaranteed the fact that they will be soon driving a Continental Mark II are a royal head of a Middle East country, an Eastern potentate who derives his income from rich oil fields, and the President of one of the larger South American Republics.

Anyone can buy one of the new Continentals providing he has the money, has a high standing in the community, (or perhaps, enough influence) and happens to be at the right place at the right time. If you cannot fulfill the above requirements, the only thing to do is to wait until a model comes up on the used-car market, and grab it.

The Continental Mark II is one of the most interesting automobiles we have had the honor to test, and this is taking into consideration some most attractive and potent European machinery. It is a car that will add great distinction to the Ford line in particular and the American industry in general.

It is built for the driver who appreciates fine design and fine engineering. The car is built to last, and despite its price it is comparatively cheap when the length of time it will be in active service is considered.

The Queen of the Classics has been reborn. ☆☆

The Safest Car in America

The Lincoln Capri is given top honors, barely nosing out Packard and Studebaker, in a survey which found not too wide a disparity in the current models

A CAR LIFE Staff Report

WALKING on ground where engineers fear to tread, CAR LIFE research experts rate the Lincoln Capri four-door sedan as the safest car in America. Reaching their conclusions on a safety-percentage basis, the researchers found a not-too-wide disparity among all 35 models reviewed.

A prodigious amount of work was involved in making the ratings, practically all of which demanded original study and the creation of an accurate scoring system.

Although the Capri stood up to detailed investigation better than any other, its contemporaries were not too far behind. Using a base of 100, the research group rated Lincoln 73 per cent. The Packards, both Clipper Custom and Patrician, tied for second place with 72 per cent.

The overall average of the 35 models reviewed, which included four-door and two-door sedans, convertibles and station wagons, was just over 62 per cent. With Lincoln 11 points above the average and the lowest rated car scoring 57 per cent, it can be seen there is little to choose from in safety value in most of the current American cars.

Much of the Lincoln's safety factors can be attributed to its excellent suspension system (with regard to weight distribution), chassis design and construction, and its fine steering layout, which is admirably suited to the general design and construction. It cannot be denied that the manufacturers have learned from the Mexican Road Race, where Lincolns have constantly figured in the winning circle, and these lessons have been passed onto the motorist—the chances of a Lincoln misbehaving itself in a tight spot are less than with any other American automobile built in 1955.

However much the automobile manufacturer tries to make a car safer to drive, his efforts can be completely nullified by an incompetent driver. The manufacturer can only do so much to protect the driver of an automobile; his provisions can only hold good so long

How They Stand

1. Lincoln Capri	73%
2. Packard Clipper Custom	72%
Packard Patrician	72%
Studebaker Speedster	72%
3. Studebaker President V8	71%
Studebaker Com. Regal V8	71%
4. Chrysler Windsor	65%
Ford Customline	65%
5. DeSoto Fireflite	64%
6. Nash Rambler	63%
Nash Ambassador	63%
Ford Country Squire	63%
Mercury Monterey	63%
7. DeSoto Firedome	62%
Ford Sunliner	62%
Oldsmobile 98 Series	62%
8. Hudson Wasp	61%
Hudson Hornet V8	61%
Cadillac Series 62	61%
Oldsmobile 88 Series	61%
Pontiac Chieftain V8	61%
9. Chrysler Imperial	60%
Dodge Custom Royal V8	60%
Buick Roadmaster	60%
Chevrolet 150 Series	60%
10. Plymouth Plaza	59%
Ford Thunderbird	59%
Cadillac Series 75	59%
Pontiac Star Chief	59%
11. Dodge V8 Convertible	58%
12. Plymouth Belvedere	57%
Buick Century	57%
Cadillac Eldorado	57%
Chevrolet 210 Series	57%
Chevrolet Bel Air	57%

as the driver uses reasonable care and is reasonably competent. In actual driving there is no such thing as a 100 per cent "safe" automobile, but a theoretically ideal "safe" car can be determined by detailed evaluation of such items as suspension design and construction, chassis layout and construction, weight distribution, engine performance, braking efficiency and many other facets of the automobile. All these factors have to be considered in relation to one another and in relation to each particular type of vehicle. A set of safety factors for a convertible are by no means the same as the set of safety factors for a four-door sedan.

Should the "ideal" car be driven by a "reasonable" capable man it can be assumed that it is less likely to become involved in an accident due to faulty design or mechanical failure than any other vehicle. Should the theoretically impossible happen and the car be subjected to crashing, the construction of the chassis and body will offer the maximum protection to the driver.

In studying the subject of the safest car manufactured in the United States, CAR LIFE's research group discovered that, to date, there has been far more investigation into what happens to an automobile after an accident than what happens before an accident. There is a lack of available data on roll factors for various cars—factors that disclose at what speed and under what conditions a model tends to roll. There is practically no published information on the redistribution of weight, and what forces act where, when an automobile is driven under abnormal conditions or cornered at top speed. These facts would disclose the inherent safety of the vehicle and the extent to which it would "misbehave" when involved in a difficult situation. A great deal of work on these matters has been carried out abroad and more study on these subjects is needed by the American manufacturer. It is trite but true to say that prevention is better than cure. It is better to *try* to construct a car that will not become involved in an accident, however abused, rather than construct a car that is moderately safe should it be in an accident. That an automobile is going to be involved in an accident should not be accepted as dogma.

CAR LIFE's researchers, after much detailed investigation and the study of reams of specifications and research data came to the unanimous conclusion that the most important safety factor in any automobile is the design and construction of the suspension system, its relation to the construction and design of the chassis and the general distribution of the various masses.

The Safest Car in America

The Lincoln Capri is rated America's safest car by CAR LIFE researchers.

From the vast amounts of accumulated figures and opinions, the research group broke down all safety factors into 11 divisions of varying importance. These 11 sections were carefully graded and weighed according to their influence upon the safety pattern of the "ideal" car, which was given a factor of 100. Once the basic evaluation was determined, the 1955 model cars were minutely compared against the "ideal" and a percentage safety factor arrived at for each car tested and studied. Each individual factor, whether it was windshield design or mechanical reliability of the engine, was dealt with from the sole point of view of that one particular model and in relation to all the other individual aspects of the specific car. It was found that in some cases what was good for one particular four-door sedan model was not necessarily a good safety feature for another four-door sedan of the same make. The cardinal factors considered in our study of the cars follow:

Suspension Design and Construction: 14 Points

In considering these facets of the automobile we took into consideration its geometrical design, rigidity of the system and its action under varying conditions. The effect of weight redistribution was closely noted as was the design and construction of the chassis and body. This division received the greatest amount of attention.

How They Rate By Basic Price Groups

GROUP 1. High-priced:
1. Lincoln Capri 73%
2. Packard Patrician 72%
3. Studebaker Speedster 72%

GROUP 2. Higher middle-priced:
1. Packard Clipper Custom 72%
2. Chrysler Windsor............ 65%
3. DeSoto Fireflite................ 64%

GROUP 3. Lower middle-priced:
1. Studebaker President...... 71%
2. Mercury Monterey.......... 63%
3. DeSoto Firedome............. 62%

GROUP 4. Low-priced:
1. Studebaker V-8................ 71%
2. Ford V-8........................... 65%
3. Nash Rambler.................. 63%

Chassis Design and Construction: 12 Points

As with the suspension system the chassis design and construction was reviewed in conjunction with weight and body design, etc. The weight of the automobile has little effect upon the actual safety of the car itself. It will be recalled that the force of impact is equivalent to $\frac{1}{2}mv^2$, where m is equal to the mass of the object and v represents its speed. The distribution of the masses, however, was either a favorable or unfavorable safety factor.

Steering Mechanism and Operation: 10 Points

In all cases the conventional type of steering was involved unless power steering was a standard fitment. The use of power steering, although making the work of the driver less tiring, cannot be assessed generally as a good safety feature, although some systems are safer than others.

Braking System and Effectiveness: 10 Points

Power brakes were only dealt with when they were standard equipment. The ability to stop in a very short distance is in itself not necessarily an overriding safety feature since there is great danger in pulling up too sharply. Positiveness and reliability are the main factors concerning this phase of the automobile. The use of certain types of automatic transmissions throws an extra burden on the braking system, and in such cases the braking system should be developed to take care of the extra work.

Mechanical Reliability of Engine: 9 Points

The effects of engine failure are all too well known, but become more so when they affect certain types of power equipment, which cease to function if the engine is not working. Considerable attention was paid to the reported frequencies of engine repairs which indicated certain weak assemblies and parts.

Body Design and Construction: 9 Points

Consideration was given to the shape of the body and the effect of airflow and pressure. It will be remembered that some of the early postwar cars were very bulbous and at certain speeds became nose-light, due to the air lifting the front fenders. The construction of the body was reviewed both from the point of view of rigidity of manufacture and the manner in which it was integrated with the chassis.

All-round Visibility: 8 Points

This was considered purely from the driver's seat, with attention being paid to blind spots. The fact that there was good all-round visibility invariably meant a large area of glass—a vulnerable and dangerous material. Good visibility had, therefore, to be weighed against any large expanse of glass.

Engine Performance: 8 Points

The ability of the engine to pull the car out of trouble was the prime consideration in this division, and torque was noted at various speeds and rpms. The study of the engine performance was with due consideration for the particular type of vehicle, its weight, braking power and structural ability to take violent changes of power. In this respect, the operation and construction of the transmission system and rear-axle assembly was also reviewed.

Transmission: 8 Points

Conventional gear-boxes and normal drives were investigated, unless automatic transmissions were standard equipment. As stated above, the transmission system was reviewed in conjunction with the engine and rear axle. Certain types of automatic transmissions that lock upon failure are a definite hazard.

Windshield Design and Construction: 7 Points

As the wraparound windshield has become an almost universal feature it was decided to add this special division. It cannot be accepted that the wraparound has increased the safety of the automobile, and its adoption is considered to be of no greater value than a sales inducement. The blind spot caused by the door post has not been eliminated but merely moved to another place. In bad weather the increased amount of unwipeable windshield is a menace. No wraparound lacks distortion, for no piece of curved glass can possibly be free from refraction. The shifting back of the door post to accommodate the wraparound has, in many cases, produced a dangerous point in the previously clear doorway.

Interior Safety Precautions: 5 Points

The facts that counted in this section depended upon the amount of attention

A well-appointed instrument panel is a Capri feature.

Smooth flowing lines, high visibility make the Capri a most attractive car.

the manufacturer had paid to such things as padded tops to the instrument panel, the design of the steering wheel and its vulnerability, the firmness of seat anchors, and other conditions that would affect the driver after an accident has taken place. The two-door sedans and convertibles started with a minus in this section since in the event of an accident there is a tendency for the front seat to go forward on impact, throwing all the weight of the rear-seat passengers on those in front.

Many other aspects and factors are included in the 11 sections of the table which are merely convenient groups from which certain conclusions can be drawn. ☆

Styling of the

A leading authority on the old Lincoln Continental takes a look at the Mark II — and comes up with some surprising conclusions

New Continental

BY GRIFF BORGESON

TODAY, scant weeks after the introduction of the Continental Mark II, the car's lasting fame already is assured. The reason is that it is one of the most controversial cars ever built. Like a serious and daring work of art—which it is—it's the sort of creation that you either strongly admire or dislike. You cannot ignore it or expect not to be moved by it, either one way or the other. Some people will see the new Continental as the ultimate in good taste and enduring beauty. Others, who have always thought of the old Lincoln Continental as being years ahead of its time, will see the new namesake as an anachronism, a throwback to the past, a conception out of phase with the spirit and form of today. But one thing that all critics must admit is that it is a quality car, the like of which has not been built in the U.S. for two decades.

The old Continental had the *look* of quality which earned it its status as a classic but beneath its sheet metal it was a modified, lower middle-class, production car. When Ford was forging the philosophy of the Continental Mk. II the planners agreed that the new prestige car must come up to the most critical standards of excellence in every way. They specified not only the appearance and substance of quality, but they placed careful emphasis on the subtle feel of quality as well. I listened to discussions of these lofty goals with sceptical interest but found it hard to see how they could be achieved in mass production, even if volume were low.

Now I've examined and driven the production Mk. II. This is my reaction:

(Continued on next page)

Griff Borgeson drives the Continental Mark II. Only traces of the original Lincoln Continental's beauty are retained.

Continental Mark II dash includes usual instruments plus a tachometer. Behind bottom of lower spoke on the steering wheel is a little gold-colored plate upon which the owner's name (or signature if he so desires) will be engraved.

BUILDING QUALITY

More time is spent in metal finishing and body painting the Mk. II than is normally required for the entire assembly of most upper-bracket Detroit cars. I have been through the Continental plant and speak from first-hand observation. The best way to describe it is to call it a modern, efficient factory specifically designed for the assembly of automobiles by hand.

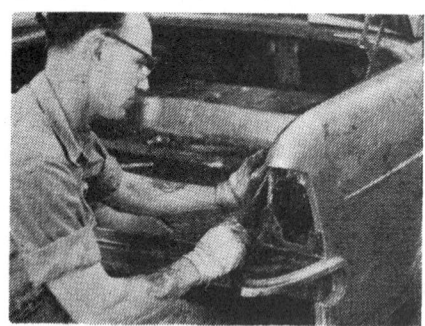

At the beginning of the assembly line (maximum rate of production is about 20 cars a day) the sheet metal body panels and supporting superstructure are assembled on a chassis frame and are adjusted to fit each other within very narrow tolerances. Any blemishes or irregularities in the metal are corrected by hand by means of skillful, painstaking filing and sanding. I saw no evidence of the use of lead or body solder. Then the panels are removed, tagged according to the frame to which they have been so carefully fitted, mounted on a big mobile jig and sent down the line. The frame is sent along another line to have its engine, drive train, and running gear installed.

The body goes through an unusually careful and thorough rust-preventive treatment and is ready for paint. No dipping, no spray booth, no mass-production methods are used. A painter stands there in the open and sprays on two coats of primer after which the paint is water-sanded by hand and then baked. Then two coats of surfacer are applied, followed by hand-sand and bake. Then two coats of lacquer color go on, followed by hand oil-sanding and bake. Then two more coats of color followed by the same hand-rubbing and then hand-polishing. The result is a rich, deep, glass-smooth finish.

From here the body goes to the trim line, alongside which is a large battery of sewing machine operators working with broadcloth as rich as the finest fur felt, brocades, nylons, and leather. This last is the finest I have seen, is as soft as chamois and flawless. The upholstery covering is installed over deep foam rubber, then instantly covered with protective sheets of plastic which remain in place until the car is delivered.

Interior accessories also are installed on the trim line. These include the motors for raising and lowering the windows. The motors are sealed in plastic bags and then the interior door space is sealed off with a plastic sheet. All this to protect the motors from moisture.

Here, too, the bank of instruments is installed. This unit is worthy of a Tiffany window; the large, honest, un-gimmicked dials are splendidly legible and are finished like costly timepieces. After being screwed into place this panel is permanently welded in, to eliminate it as a possible source of future squeaks.

The body goes from the trim line to be reunited with its original frame, now a perfect, dynamometer-tested and tuned chassis. Then comes a shakedown on the plant's own test track to pinpoint possible bugs in performance and body assembly. Finally the car is given an absolutely ruthless screening by a quality control inspector whose job it is to see that no flaw exists in the car's finish, inside and out. Having successfully run this gauntlet the car is buttoned into a tailored white cloth jacket to await shipment. When shipped, its perfection is protected by a huge plastic bag in which the car is sealed. Little is left to chance.

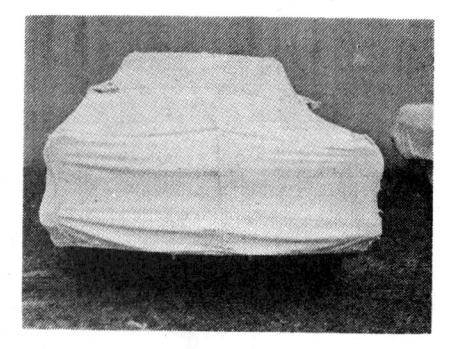

Dealer preparation of a Mk. II merely consists of mounting the wheel discs which are packed in the car's luggage compartment during shipment.

THE FEEL OF QUALITY

The car has the integrity, the "oneness" that the designers said was their aim when they formulated what they call the Continental Concept. The car has the zero-tolerance feel of a bank-vault door or a fine turret lathe. All sloppiness, play, and sluggish response to controls have been designed out of it. The steering, like the rest of the car, has a rock-solid feel. It is perfectly positive in its action and is the most vibration-less steering mechanism I've ever experienced. The swinging tail light which covers the fuel filler cap has this smooth, machined, feel of absolute rightness. The lids of the ash recepticles have it: it's everywhere. Writers of ad copy will be sorry they used up the superlatives before this connoisseur's car came along.

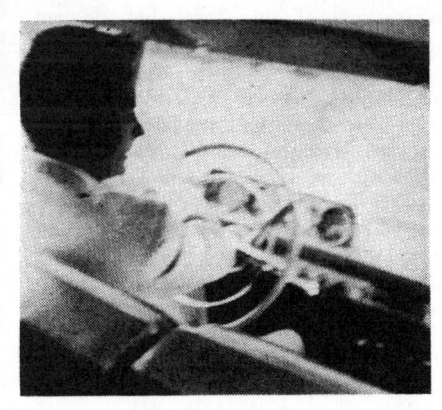

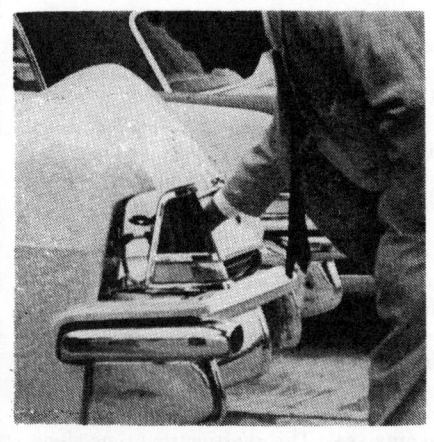

While driving the Mk. II I did not experience the airborne, riding-on-a-creampuff sensation that is a trait of many luxury cars. Instead, I felt that I was in a great automobile, firmly cushioned, solidly supported, very safely packaged and in full command of the machine and the road.

As the designers say, the car's power is entirely "adequate." There's nothing fierce about its acceleration but the Mk. II is as quick as it needs to be. During my short shakedown cruise on the half-mile test track the car seemed to handle extremely well in the turns and I found that the power brakes could be applied in a disciplined manner—were not either all-off or all-on, as they are in some other cars. The car just feels perfect, that's all.

THE QUALITY MARKET

The Mk. II is *not* a '56 Continental; annual facelifts and frequent model changes are not contemplated. If you've created something good, why change it for the sake of novelty alone? There's no real competition right now to hasten the Mk. II's obsolescence and it can exist aloof from the vicious circle of annual model facelift and change. At least that's the present plan and it works well for the very few European cars in the Mk. II's exclusive class—cars which also cleave to traditional styling, by the way.

For the average citizen, and for the present, Mk. II's are likely to be very hard to get. Civic leaders, respected public figures, celebrities of the more solid sort were the by-invitation-only guests at large but very exclusive preview showings of the car held recently in five major U.S. cities. If your name's not in "Who's Who" or in "Poor's Directory of Executives" you might have to wait a long time for delivery. The company the Mk. II keeps is not being left to chance.

THE SYMBOL OF QUALITY

In view of the remarkable amount of attention to minute detail lavished upon the new Continental the choice of an insignia for the car is surprising. The four-pointed star, used in two different forms as a hood and rear-deck decoration, is a commonplace device in industrial design and currently is in use on products of each of the Big Three. On top of that, Lincoln's recent adoption of the four-pointed star with four smaller points radiating from the center seems to further weaken the chances of this symbol becoming associated in the popular mind with the Continental and therefore with quality, good taste, and the other ideals the car clearly stands for.

THE FORM OF QUALITY

Partly because of its tasteful but high-spirited elegance, but mostly because of its name, most people have assumed that the old Continental drew its inspiration from advanced European design. I accepted this implication until I saw the first clay mockup of the new car about a year and a half ago. At this point a somewhat radical idea occurred to me, one which has been borne out by careful investigation. The "continental configuration," I am convinced, actually is a distinct *body style*, about as rigidly defined as coupe, sedan, or roadster. The formula consists of a combination: low profile; long hood; compact passenger space; blind rear top quarter; short, chunky trunk or luggage compartment; visibly-mounted spare tire at the rear.

You can easily trace the continental formula for decades. In 1921 Jordan had a continental-type coupe. In 1925 Brooks Quinsler designed a body on the Franklin chassis for showing at the New York Salon in December of that year. It had all the "continental" hallmarks and may be regarded as a 30 year-old forerunner of the present Continental. Franklin's "continental" had lots of *chic* for its time, was enthusiastically received by the public, and went into mass production. I'm

THE PRICE OF QUALITY

At this writing the Mk. II's price still has not been announced but the general feeling is that it will fall between $9,000 and $10,000. Although such a price is unprecedented for Detroit products, this is an unprecedented car. That much quality is there for those who value it and that much distinction is there for those whose only recourse has been to "buy foreign." Since we're enjoying the highest living standard in human history I think the chances are good that 4,000 Mk. II units might be sold during the first year of production. Assuming the above price, this would add up to a gross value of between $36 million and $40 million. Since we are told that the investment in the new Continental is $20 million, that would prove that there's still a very important place for quality design and workmanship in the midst of the age of automation.

Comparison of Mark I (1948 model) and Mark II rear ends. Edsel Ford and his stylists conceived beautifully harmonious contours difficult to recapture.

Body builder LeBaron, too, picked up the graceful continental styling and employed it here on a 1931 Packard chassis. Note the blind rear top quarters.

Continental form in motor car applications became more fixed with Brooks Quinsler's Franklin coupe, shown at New York in 1925, prior to production.

Lincoln Continental Number One—Edsel Ford's personal car—still had Lincoln Zephyr hub caps in this photo. Today, it remains a striking, ageless-looking car.

not aware of specific earlier European expressions of this form but do not doubt that they existed.

In the late twenties an American coachbuilder named Waterhouse began to make a specialty of this body style and Rollston, LeBaron, and Weyman also picked it up. Saoutchik in Paris, Carrozzeria Touring in Milan, and Pinin Farina in Turin were among many European coachbuilders who also were building handsome bodies to this formula. Then, for the 1931 New York Salon, Waterhouse provided the body of the Stutz DV32 *Continental Coupe*. This is the earliest example I have of the coupling of the name with the body style.

Through the first half of the thirties coachbuilt continental bodies were used often on Packard, Stutz, and Lincoln chassis, and the style's popularity grew in Europe. When Edsel Ford—who had been in charge of Lincoln coachwork for years—decided to have a car built on the Zephyr chassis for his personal use. he chose the functional, beautiful, and by then well-established continental form. When it became a production car it was given the same name that Stutz had used eight years earlier. Most critics considered it a design masterpiece. So much for the past.

The Continental Mk. II represents a carefully calculated effort to design a modern masterpiece based upon one of the most elegant body styles of the classic period. It will be fascinating to see how the public reacts to this "traditional" approach. There is little doubt that 2,000 to 4,000 Mk. II's will be sold during the coming year. The question that time alone can answer is whether the new car's form can have the same impact on the public that its honored ancestor had. Will it be accepted as a masterpiece? Its rich but austere beauty can't be judged by the photos on these pages alone. You have to be in its physical presence.

The fact that this body style is a traditional configuration seems to me to eliminate controversy over the method of mounting the Mk. II's spare tire. It *had* to be mounted at least semi-visibly because it had been one of the most important hallmarks of the form. Originally the visible spare added to the functional integrity of the design. It was so admired that "continental kits" flooded the market and the designers of the new Continental were forced to take a different approach, forced to sacrifice the fact of function in order to maintain the form of function.

The solution they adopted works very well. It is very easy to remove or replace the spare, as it is located in the new car. There is considerable room on either side of the spare for passing objects into the luggage compartment. And it is a handier, more useful compartment than the one on the old Continental. •

Ford Motor Company

Position

1. Lincoln Capri
Four-Door Sedan

Safety Rating
(Percentage)

73

The Safest Car In America!

THE LINCOLN CAPRI has the distinction of both theoretically and practically being the safest automobile out of the 35 we have covered in these pages.

The Capri's suspension and chassis construction and attendant good steering give it a high degree of stability on the road. Under adverse conditions the car sits firmly on its four wheels without any pronounced inclination to roll. Redistribution of weight has been accounted for in an excellent manner. Rear wheel shudder on a bumpy road is kept to a minimum.

The design of the body of the car and its effect upon the chassis contributes a great deal to the good overall figures.

The performance of the engine is well in keeping with the car's other characteristics and its reliability is among the best.

The structure of the Capri in general is good, with its body of a high tensile strength offering protection in case of an accident.

The Capri is the result of successful repeated participation in the Mexican Road Race, where cars are forced to their absolute limits. A car that comes through that gruelling contest, and wins as the Capri has done, cannot but have a basically high safety factor. What has been learned on the roads in Mexico, under the most exaggerated conditions, has been passed onto the every-day motorist who should, with a 1955 Lincoln, have confidence in the car he is driving.

Engineering of the New Continental

BY ROGER HUNTINGTON

The chassis department is one area where the car really is different. So here's a top expert's analysis, plus some predictions on performance

ONE OF my first impressions of Ford Motor Company's new Continental was that they're going to lose a wad of money on every one they build!

This is a *different* kind of car. We haven't done anything like this in the American auto industry since the old 12- and 16-cylinder super-luxury boats of the 1930's. Sure, we have our expensive "prestige" cars today—Cadillac, Packard, Lincoln, Imperial—but these are out-and-out *production* cars. You can find a thousand places on any one of them where quality has been compromised to save a dime. This is okay . . . it's the only way they can make a decent profit in today's market.

The Continental, on the other hand, is one of the very few cars being built in the world today without this heavy concentration on cost. I won't say there aren't *any* compromises on it —there have to be compromises on any practical engineering design—but I'm convinced that none of the basic objectives of the design were purposely compromised to save cost. I'm also convinced, after seeing the Continental being hand-built, that they would have to sell this rig

Test bench check on Mark II V-8 which is standard '56 Lincoln except for air cleaner, oil pan, special exhaust manifolds to clear chassis, and the finned rocker covers.

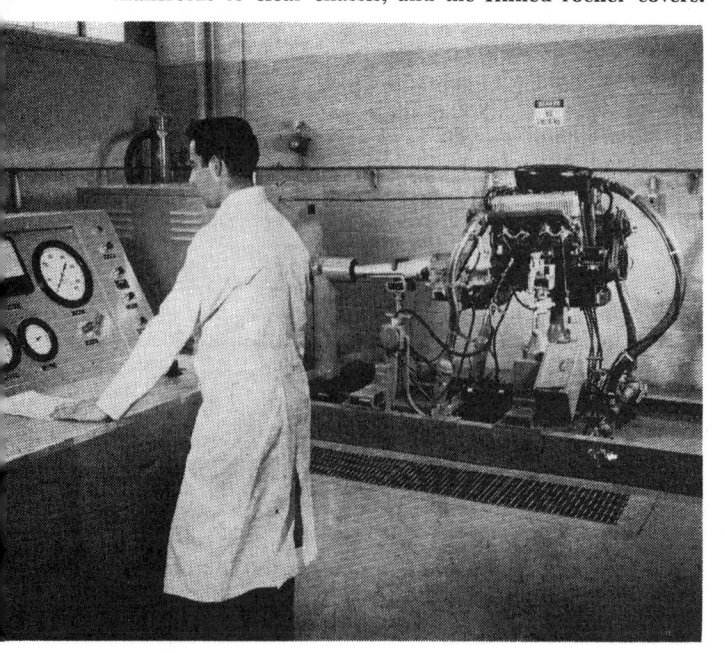

Special tubular frame crossmember passes under the engine on the Continental. Unique oil pan is necessary to clear it and allow a reduction in overall engine height.

Continental chassis with running gear. The frame is a brutal thing, with seven tubular cross members and a massive backbone running down the center. It weighs 530 lbs.—or about twice the total of a conventional X-member frame. As a matter of fact, the Continental's frame has more torsion and beam stiffness by far than anything else.

Distance from axle to rear spring front pivot point is shorter than the rear distance; this shortens the effective lever arm for spring windup torque, eliminates twisting.

Driveshaft has double U-joint to permit routing of the shaft above the frame's crossmembers. The practice is not new, but is used rarely in contemporary American autos.

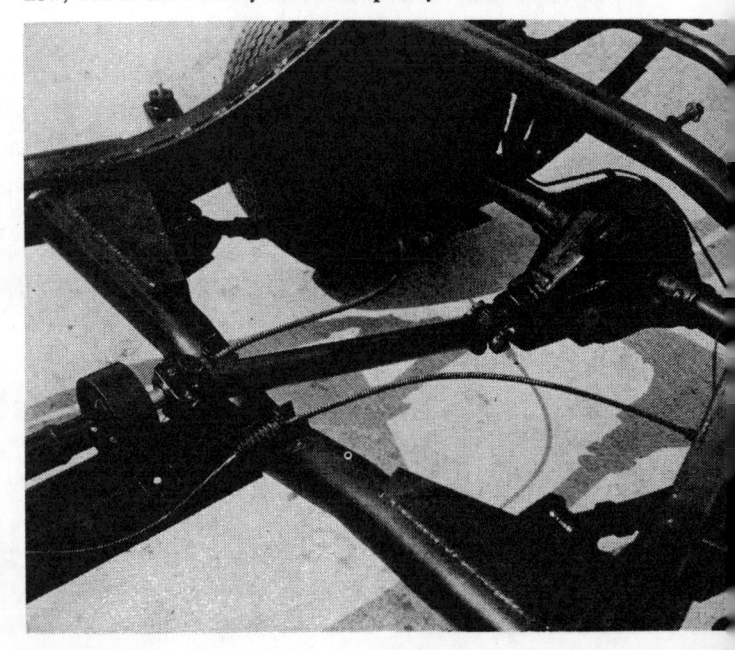

for at least $15,000 to show a decent profit! It's that kind of a car.

Now before we get into a technical analysis of this brilliant design I want to make it clear just what these "basic objectives" of the design were—and what they *were not*. There were several primary objectives: (1) Striking, luxurious appearance and "feel" inside and out; (2) best possible ride on all types of road; (3) a "solid" feel to the driver; (4) maximum silence at all speeds; (5) a body that will stay tight and solid for several years. They were *not* concerned about (1) world-beating performance, (2) light weight, or (3) a car that would last for 20 years.

In other words, Ford is after a car they can build in very small quantities that will take prestige leadership away from Cadillac. This is considered an important factor in Ford's overall plans for future expansion in the automotive field (in line with the coming public sale of stock). They know they can't oust Cadillac with just a price tag and good looks. They can't hope to do it with brute road performance, what with the high weight of the car and the rapidly increasing performance of cars in all price classes.

Harley Copp, head Continental engineer, is especially proud of frame which is unique with right angle bend in front.

Continental passengers each will sit within boxed section of frame, a design likely to be imitated by less expensive makes.

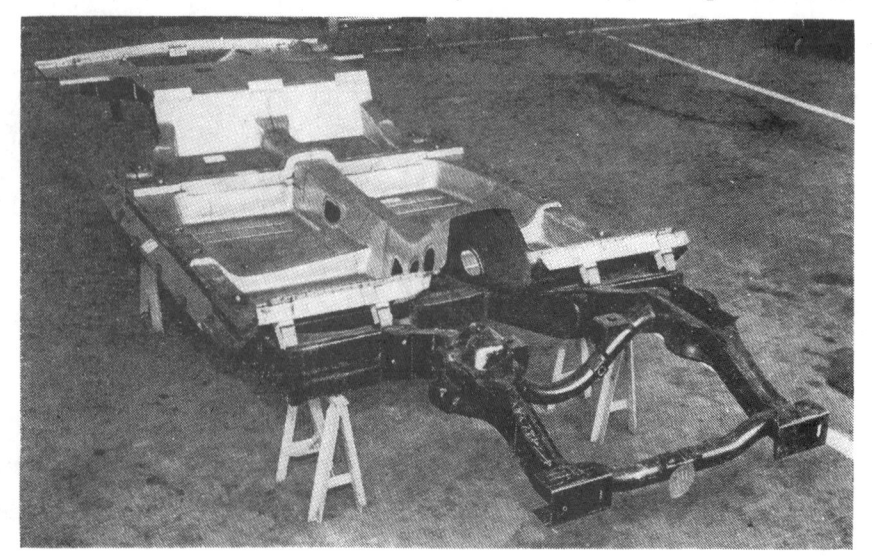

They feel that a car that *rides* and *drives* better than anything we know today may do it. They have such a car in the new Continental.

Here's how they did it.

I think the frame is the heart of the new Continental design. It had to be a lot different than anything we know around Detroit now. In the first place, the styling boys called for an overall car height of 56 inches, or a drop of about five inches from current bodies—without any great loss of headroom and seat height. This meant a much lower floor. You just can't get a lower floor with a conventional X-member frame without greatly reducing road clearance, which of course is out of the question. The answer was a "ladder" type of frame with straight right-angle cross members. Then you use "footwells" that drop down between the frame members for the floor. (They use this gimmick on the Studebaker coupe.) This permits the floor level to be only about eight inches above the ground!

But there's a bug: A great disadvantage of the ladder-type frame is that it lacks torsional stiffness—that is, it doesn't have much resistance to *twisting* when passing over bumps and dips in the road. The Continental engineers couldn't have this; not only would the twisting upset suspension and steering geometry, but you couldn't keep the body from squeaking. Terrific torsional stiffness of both frame and body was a must. They got it in the frame by using boxed side members, tubular cross members, and a massive "backbone" running down the center and V'd into the front end supports. The new frame is a fantastic thing to see. It weighs 530 pounds, or about *twice* the weight of a conventional X frame—and it packs far more torsional stiffness than anything in the industry. Incidentally, the straight cross members and backbone forced a two-section drive shaft. Money, money, money . . . and more and more engineering problems!

Most of the front and rear suspension units are standard 1956 Lincoln. But there are some subtle differences that change the character of the car completely.

Take rubber cushioning, for instance. We usually think of rubber as being just about the perfect cushioning medium. Actually, it's *too* elastic to suit the Continental engineers. In other words, there's very little damping effect when it absorbs energy, and you get a healthy energy "kick-back" on the rebound. On the front suspension they have done away with the rubber rebound bumpers and replaced them by progressive hydraulic restriction in the double-acting shock absorbers. Near the rebound end of the shock piston travel they have three orifices of progressively smaller diameter through which the oil escapes; the piston cuts off one orifice after another—which causes the damping resistance to shoot up—and finally you get into hydraulic lock (no flow) for the final stop. It's all silk-smooth, and there's no kick-back. You never get into this effect on a smooth boulevard, but it makes a world of difference on dips and bumps.

Another important feature of the new Continental shock absorbers is the thermostatically controlled main valve. With an ordinary shock, when traveling over a rough road, the oil in the shock heats up from the violent turbulence, thins out, and you lose much of your damping action. The ride suffers. On this new design a tiny bimetal thermostat varies the tension on the valve spring as the temperature of the oil changes—maintaining constant shock action under all conditions. Works like a charm.

They also have gotten rid of the rubber bumpers above the front half of the rear leaf springs, which are used on the standard Lincoln to cushion spring "wind-up" on hard acceleration. This was never a very good deal. On the Continental they can get along without the bumpers by shortening the distance from the front spring pivot to the axle (and making it longer behind); this shortens the effective lever arm for wind-up torque, and practically eliminates the spring twisting.

Another vital spot where Continental engineers have eliminated rubber is in the suspension bushings. We've been using rubber bushings on spring shackles and A-frame pivot joints for years because they were very cheap and would function without lubrication. But the elasticity of the rubber also takes a lot of that solid feel out of the handling. On the Continental they use threaded metal bushings. Expensive . . . but wait 'till you feel that steering!

In fact, it should be obvious by now that Continental engineers have spared no expense to make this rig ride and handle like a Rolls-Royce. Other contributing factors to the fine ride are the high curb weight of 5,000 lbs. and the greater-than-average proportion of weight on the rear end (46.4 per cent)—achieved by using the long hood and setting the engine back a little. Well, you just have to ride in the car to appreciate it.

Continental stylists have achieved certainly one of the most appealing blends of classic and modern body design to be seen in the world today. But, just as important, the slide rule boys have made that body tighter and stiffer than anything we've known in America since the days of the classic car. The body is absolutely silent on a rough road—no squeaking, drumming, no vibrating panels. And there won't be any sagging doors and loose fits showing up after a year or two either. Torsion tests on the new Continental hardtop body show considerably more rigidity than any current Detroit body, even the turret tops.

How did they do it? Probably the most important contributing factor to body stiffness is the structure at the front (by front I mean around the dash line, which is the front of the body proper). Conventional bodies that are used with frames have the bottom side rails ending just below the front window post, with a vertical column going up to the post. On the Continental they bring these rails forward another 16 inches or so, then use massive *diagonal* bracing from the ends of the rails up to the dash—in addition to the regular vertical pillars. Not only does this add tremendously to the beam and torsional stiffness of the body as a whole, but it makes that front vertical pillar that carries the door a lot more rigid. You *must* have this post very solid if you want to prevent squeaks and loose door fits developing over a period of time. Further stiffening in this area is provided by welding in the instrument panel instead of bolting.

An obscure advantage of having terrific stiffness at the dash section of the body is the reduction of annoying steering wheel shake on a rough road. With a non-rigid dash structure and long steering column extending out from the dash anchor point, it's murder. With the deeply "dished" safety wheel on the Continental (also on all 1956 Ford products) there's practically no column at all above the dash . . . and that wheel is solid as a rock on any kind of road. Great engineering!

One especially tough body problem was that the more or less flat roof the stylists wanted to use tended to "drum." They solved it by using three cross bows instead of the usual one. And just as an example of the lengths to which the Continental engineers went to reduce noise: The front wrap-around windshield and side windows are "countersunk" in the upright front corner posts to present a flush, streamlined surface to the wind passing around the corner. Other cars have a large bulge here which is the cause of a lot of wind noise at high speed. The construction is very expensive . . . but that's what I mean about this new Continental.

UNDER THE HOOD

The Continental uses a standard 1956 Lincoln engine and Turbo-Drive torque converter transmission. Only differences are a special air cleaner and oil pan to reduce the overall height of the engine, special exhaust manifolds to clear chassis parts, and finned rocker covers to give the engine a distinctive look. Contrary to rumor, they're not using dual quad carburetors to boost the horsepower rating to 300. Matter of fact, Continental sales people are not talking horsepower at all. No certified rating will be published. They're tearing a leaf from the Rolls-Royce book and just saying their car has "adequate power." They figure anybody who can pay $10,000 for a car isn't going to quibble about five or 10 hp. (For comparative purposes, of course, you can just consider the Continental as carrying the Lincoln ratings—285 hp, 401 lb.-ft. of torque, 9:1 compression ratio, etc.)

I would like to emphasize at this point that the engines that go into the Continentals are not quite like the ones that come off the Lincoln assembly line. Each and every one is exhaustively run-in and tested on the dynamometer before it goes into the car. Power and torque output, carburetor metering, spark advance, etc., must all fall within very narrow limits. Every engine is *right* before it goes to an owner—and there's no lengthy breaking-in period either.

The dual exhaust system is interesting. There wasn't any space for it *under* the chassis, what with the footwells and tubular cross members; so they ran the pipes along the frame side members, set back into indentations pressed in the channels. The only place they could put the mufflers was at the extreme rear of the lines. This is a blessing in a way, since the exhaust gases have cooled considerably by the time they get to the mufflers, and they can get by with quite small ones. Incidentally, you won't get the usual "blat" under load you get with duals on a V-8 engine. There is a neat cross pipe connecting the two exhaust lines just ahead of the mufflers; this will balance the exhaust pulsations and give a smoother hum.

GENERAL COMMENTS

I don't consider myself a connoisseur of auto styling, so I'll keep my comments on the esthetics of the Continental design to a minimum. I did want to mention the beautiful instrument panel. Continental public relations people tell me that their designers and stylists were guided in many difficult spots by the results of customer questionnaires that were circulated many months ago. The instrument panel was especially tough. The people who answered the questionnaire wanted a clean, simple panel—but when asked what specific instruments they wanted included, they wrote down everything from sump temperature gages to altimeters! They've compromised on a neat layout of the usual five instruments—speedo, oil pressure, fuel, amps, and water temp—plus a tachometer and sweep-second clock. It's beautiful, and all gages are of the sweep-needle type. No flashing lights to tell when the oil pressure is low. Controls for the air-conditioning are on a sub-panel stretching down from the dash to transmission bulge.

There are only two sour spots in the whole design as far as I'm concerned. The very low body, swept-back front windshield post, front door pillar, and "step-down" floor combine to make it a hard car to get in and out of. This was a hard compromise the engineers had to accept in the name of style. And that upright spare tire in the trunk leaves me cold. It's strictly a non-functional arrangement. The upright spare does nothing but choke off access to the trunk compartment. The bulge in the lid is an interesting styling gimmick, but it looks phony as a three dollar bill on a modern body. The Continental engineers said they just about had to retain a conspicuous spare tire to tie the design in with the old Continental. Could be . . . but it's my prediction that they'll be taking that upright spare out of the deal one of these days.

Now I know you auto enthusiasts are wondering about performance. Naturally we haven't had a chance to clock any actual test figures yet, but I think I can hit fairly close on the slide rule. The high curb weight of 5,000 lbs., torque converter transmission, and 3.07:1 axle ratio will combine to limit acceleration. Furthermore, even though 368 cubic inches speak with authority, you never know whether those 285 rated horses are big, hairy stallions or ponies. The top speed of the 1955 Lincoln indicated no unseen horses. I'd say your 0-30 mph time on the Continental will be four to five seconds, which won't beat the pack from the traffic light any more. It should get to 60 in maybe 11½ to 12 seconds, which will match or beat most '55s (don't know about '56). With all the inches and potential horsepower, plus the 3.07:1 rear end, top speed should be 'way up there. Actually, I'd be surprised if it breaks 110 mph. The "cool" Lincoln valve timing is great for low-end torque, but it just doesn't pull big on the top end.

And guess that about covers it. . . . •

'56 LINCOLN ROAD TEST

The luxury field's only brand-new car shows its mettle in MT's 1st road test for '56

AN MT RESEARCH REPORT

THE BEST WAY TO START a discussion of MT's 1st '56 test car, a new Lincoln, is to acquaint new and old readers with the things we said about last year's model, the "new era" Lincoln with features that hinted at what to expect in '56.

Aside from a performance analysis, we editorialized about the car's future—namely the '56 models—and came up with the thought that Turbo-Drive would be responsible for a complete shakeup of Lincoln sales outlook (i.e., styling, ride). To quote from the '55 report (May issue), "It reflects challenge to be projected by '56 models."

The challenge? A bid for new recognition in the prestige car field. Let's see how the longer, lower '56 Lincoln meets this challenge.

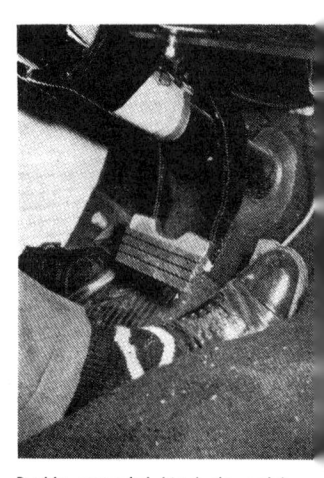

Test car: Capri 2-door hardtop, formerly most deluxe, now just a deluxe, but less expensive than the luxurious Premiere. Test car comfort-and-convenience extras included Turbo-Drive, power steering (both standard on all Lincolns), power brakes, radio, heater, and easy-does-it items like pushbutton antenna adjuster and a radio that went after distant stations with bloodhound determination.

Engine: Considerably more than just a warmed-over '55 engine, the 368-cubic-inch Lincoln V8 gets its authoritative punch from a brand-new block, new intake manifold, and husky, well-balanced lower-end components. Compression ratio, 8.5 last year, is now 9.0 to 1.

Inside the new block, pistons stroke 3.66 inches thru 4-inch cylinders with a slight increase (240 feet per second) in piston speed over the '55 engine. Peak horsepower rpms are up due to a higher-lift camshaft, stiffer valve springs, and a fully redesigned intake manifold system. Dual exhausts are standard.

Other options: Let your payments be your guide, for available options can create a lengthy list of digits on your sales order. Air conditioning, of course, is the costliest. You won't be bothered with many lesser options, for some, including windshield washers, power steering and power brakes, are "free." Body options include

Doubly suspended big brake pedal suits itself to all driving techniques. Throttle is hanging, too

112

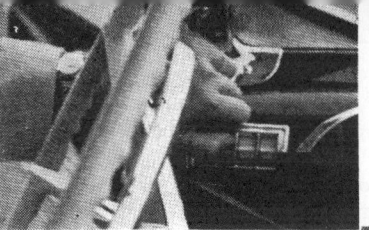

Above: Lincoln's new placement of pushbutton window regulators is not just a novelty. You can control all main windows in the car with only the slightest movement of your fingers from the wheel. Left: Extensive road test equipment almost vanishes in Lincoln's capacious trunk. Built-in exhausts have come to be expected, but rear "grille" is unique

You'll know right away that you're in the front compartment of a Lincoln, but there have been lots of changes. Wheel is dished as on all Ford products, heater controls (left, below instruments) announce what they're up to

a Capri 4-door sedan (besides the 2-door hardtop), and 3 models in the Premiere line: coupe, sedan, and convertible. Lincoln's 4-way power seat and power windows are standard in the Premiere, optional in the Capri; and (here's an interesting offer) all-leather trim is available at no extra charge. Seat belts, tho, cost extra in either series.

WHAT THE CAR IS LIKE TO DRIVE

Exit and entry: For a large car, and one with obvious space to spare, Lincoln reveals a complication—at least to long-legged drivers or passengers getting into the front seat—and that's the protruding corner of the wraparound windshield. One test driver (just over 6 feet tall) bumped his knee "too often" before realizing you can't be in a careless hurry getting in or out of the car. But the error didn't repeat after a day on the road with many driver changes. The hardtop's wide single door gives more-than-adequate entrance space, and getting into the Lincoln coupe's rear seat is very little harder than stepping into a 4-door sedan. Bright courtesy lights mounted at the extreme edges of the instrument panel go on when the door is opened, are positioned to be of real value.

Herculean V8 is new and bigger, has detail features shown by arrows. Left, new vacuum spark advance control for distributor. Center, new choke mechanism, integral with carburetor. Right, new location for air inlet and oil filter means easier servicing

Altho the '56 Lincoln is more than 2½ inches lower at the roofline than in '55, the seats are low enough to save your scalp as you settle into the car. The low seat also keeps the steering wheel out of the driver's lap as he slides under it.

Driving position: Very much different from '55. In Lincolns gone by, you sat high, with the wheel fairly low and close to you, and the cowl and hood were things to look down upon. Now, the seat is low, slanted backward just a little too much for

---

TEST CAR AT A GLANCE

'56 Lincoln Capri

ACCELERATION
From Standing Start
0-30 mph **4.0**   0-60 mph **11.7**
Quarter-mile **18.2** and **78** mph
Passing Speeds
30-50 mph **4.8**   50-80 mph **10.3**

TOP SPEED
Fastest run **106.9**   Slowest **104.7**
Average of 4 runs **105.5**

FUEL CONSUMPTION
Steady Speeds
20.2 mpg @ 30   18.1 mpg @ 45
15.5 mpg @ 60   13.0 mpg @ 75
Stop-and-Go Driving
11.0 mpg over measured course
13.6 mpg tank average for 536 miles

STOPPING DISTANCE
172 feet from 60 mph

SPEEDOMETER ERROR
Was correct at 30 and 45 mph, but read 58 at true 60, 74 at 75, and 115 at top speed

REAR WHEEL HORSEPOWER
Clayton chassis dynamometer showed:
87 road hp @ 2000 rpm and 46 mph
104 road hp @ 2500 rpm and 70 mph
116 road hp (max.) @ 3100 rpm and 80 mph

MT's drivers, and the cowl and hood take on a new appearance. (A 4-way power seat control alleviates this condition.) You're more within this car than before, and the feeling is one of a really big car. And it's just that. The short hood is a thing of the past, yet forward vision remains good, with front fenders visible to any-sized driver.

Instrument legibility is slightly better than previously, mostly because of slightly contrasting background of gold behind white letters, and red needles. Handgrips on inset wheel are positioned comfortably for long trips, but considerable glare was noted from chromed steering wheel spokes.

Vision: With glass area of this proportion, it couldn't be anything but good. Some windshield distortion noticed from the driving position near cornerposts. Posts themselves do not cause blindspots.

It's our thought that Lincoln could well go to a dashboard-mounted rear-view mirror. Our test car had a handy, knob-adjusted inside mirror that was easily inched up or down at the turn of a knob. But it wasn't enough to provide a long-distance view of the road behind. (This foreshortened view is caused by the very low streamlined roofline at the top of the rear window.) The rear window is wide, and its cornerposts don't restrict vision in any way. The inside mirror could be wider to take advantage of the expanse of glass at the rear.

Operation of accessories: Heat and vent controls are similar to Mercury aircraft-type control knobs with a slightly different twist this year: a roll-over indicator moves upward as you pull a knob downward. Heat range is shown by a red indicator filling more and more of a telltale window as you advance the knob.

Other controls (windshield wiper, lights, instrument lighting, rheostat, heat-defroster blower, etc.) are either toggle switches (a notable improvement over push-pull or turn-type controls) or conventional knobs, depending on the action they perform.

Turbo-Drive's quadrant (PARK, REVERSE, NEUTRAL, DRIVE, LOW from left to right) is well marked, with each range within a red circle as you engage it. The shift lever itself is positioned perfectly for a fast, fingertip shift without taking your hand from the steering wheel.

Ease of handling: There's nothing at all awkward or hard about maneuvering the new Lincoln, but let's face it—this year's car is longer and wider (by 7.2 and 2.5 inches, respectively) and you sit deeper inside it, so it's not going to be easier to handle than the compact '52 to '54 models, or even the slightly stretched '55.

A little more than 3¾ turns stop the front wheels at their extremes, enough to make parking the test car fairly easy; steering is still quick enough to make the car respond satisfactorily when you have to move it in a hurry at high speeds.

Pedal positions are good, with a wide, double-shaft, suspended brake pedal that's as convenient for left-footers as right-foot pivoters. With seat forward, upper leg may hit steering wheel when you apply brakes with your right foot.

Acceleration: It's interesting to note the manner in which Lincoln's new power for '56 has been applied. (It's easy to see if you have last year's test handy; if not, here's a quick rundown on what has happened): 0-30 time reduced by 0.2 seconds, 0-60 time lowered by 0.7, 10-30 and 30-50 times each better by 0.5 seconds; the time it takes to get from 50 to 80 mph has been lowered from last year's 13.7 seconds to 10.3 seconds. The pattern is obvious, particularly when you find that top speed is upped by only 1.4 mph over last year's 104.1.

Lincoln engineers deserve a pat on the back for gaining more performance and better fuel economy at the same time. With an improvement of over 3 seconds in 50 to 80 mph time and improved economy, there should be little complaint on what leadfooters might consider a not-too-high top speed.

MT's testers tried to outsmart Lincoln's Turbo-Drive transmission by shifting it at different points from LOW to DRIVE and back again, but the unit always came up with an equally good figure when left to shift for itself. In standing-start ¼-mile runs, for instance, the Lincoln's time in normal DRIVE range (with the throttle floored to engage the automatic low-gear start feature) averaged out to 18.6 seconds, with a true speed of 78 mph. (In DRIVE, Turbo-Drive shifts automatically from its 1st gear at 38 mph and from intermediate range to high at 75 mph.)

The nearest we could come to bettering this efficient pattern was when we held the unit in LOW range (manually) up to 43 or 45 mph. This shift pattern consistently resulted in slower speeds (74 mph average at the end of the quarter), but dropped our elapsed time down to 18.2 seconds. (See "Time vs. Speed," Sept. MT.)

Braking: The 1st of the '56 test cars subjected to our brake-fade test (12 consecutive hard-but-not-panic stops from 60 mph), the new Lincoln came thru with flying colors. Brake setup on this car was particularly suited to non-lockup stops, for the pedal action was such that there was no "grab" (that delicate point just short of wheel lock) experienced. Four complete stops were made with no feeling of fade or uneven pull; 5th stop required a great deal more pedal pressure, but car was brought to a complete stop in about ½-car-length greater distance than earlier stops. It was on this stop that the car 1st showed a tendency to veer to one side.

Uneven stops (pulled 1st to left, then to right as brakes were hit) continued, but Lincoln retained stopping ability right up to 12th and final stop. (Pedal pressure was ineffective until within 2 inches of floorboard at this point.) From 7th stop on, fade did not increase appreciably until 10th stop.

Swerving to one side or the other continued even when the brakes were hit after up to 5 miles of 50 mph driving. But Lincoln never lost its ability to stop. Car was exceptionally free from nose-dive in all brake tests. In a 60-mph panic stop (brakes full-on, just short of locked wheels) after brakes had completely returned to normal, the heavier '56 Capri stopped in about ⅓-car-length greater distance than last year's test car.

Roadability: Outlining his impressions of the '56 (as compared to the '55) in a Drivescription in the October issue, MT's Editor, Walt Woron, questioned the ability of the '56 to match the '55 on a roadability basis. Now, after a complete road test, we are able to pin down Walt's feelings about its handling.

He summed it up in one paragraph: "There seemed to be slightly more body lean when taking corners, but it is not uncomfortable. At high speeds the car is stable and sure footed."

A large part of the Lincoln roadability story concerns itself with "seat-of-the-pants" analysis. The new car is not the short-hooded, slight-overhang machine of the past; its bulk discourages the pleasure of literally "throwing it around" as everyone was prone to do with the smaller car of last year and before.

Mechanically, softer-sprung '56 is basically familiar, yet soft enough to be a stranger to anyone who puts many miles on the older versions. What difference does it make on the road? Actually very little, for the '56 is a true road-hugger that's not disturbed by curves banked wrong or by running right wheels onto a low, dirt shoulder at 60 mph.

There's considerably less steering wheel movement on straight and curving roads with rough surfaces; conversely, there's more steering wheel correction on straight roads than in previous models.

Directional stability rates high in every respect but one: We noticed more tendency for the '56 to be disturbed by crosswinds than former Lincolns.

Ride: Not entirely soft and pillowy, '56 Lincoln nevertheless allows a very restful ride; it's a good example of how a car can retain a degree of solid roadability and still provide good passenger comfort.

Seats are on firm side, very comfortable, and support you well enough to make long-distance drives a pleasure instead of the backache they can be in a too-soft car.

There's no unwarranted bouncing, regardless of road surface; vibration thru body is nonexistent, and road noise doesn't penetrate Lincoln's soundproofing. Even under rigorous conditions, sidesway remains low. Recovery thru dips is excellent; initial rebound is swift, but not jarring; there's no bothersome oscillation coming out of a dip or a hump in the road.

WHAT THE CAR IS LIKE TO LIVE WITH

Riding in the front seat: Because legroom is so plentiful, we weren't too concerned with measurements during the test, but a later check revealed ½-inch more legroom, plus an increase in seat adjustment travel of nearly an inch. Front seat rose ½-inch in '55, but now rises only half that far.

Riding in the rear seat: Because our test car was a hardtop coupe, it's hard to evaluate the car on a sedan-for-sedan basis. But statistically, here are some changes from '55: Hiproom is more than 1½-inch greater, shoulder room is just over an inch wider. Legroom is minutely less, and the rear seat depth (from the front edge of the seat to the seatback) is shortened one inch.

ECONOMY AND EASE OF MAINTENANCE

Fuel economy: If you're like a lot of gas-station philosophers, you'll say Lincoln owners aren't overly concerned about economy. But think of the people who consider a Lincoln "it," and who count their money to get a '56 and *are* concerned with operating costs. They'll want economy —and they'll get it to a greater degree than they did in '55.

Increases over the '55 test car range from 0.8 mpg at 75 mph to 1.7 mpg at 30. Traffic checks on MT's simulated traffic course dropped off this year, but "driving around" mileage (tank average, excluding performance tests) is impressively better.

Is the car well put together? For the 1st time in many test seasons, we had a car that we didn't have to alibi as an early production model. This test car was early production, but its finish bordered on end-of-the-year detailing. All components fitted well; exterior paint was free of imperfections. Interior trim was good appearing, altho close inspection revealed a not-the-best paint job around the door moldings. Upholstery looked neat, no frayed ends apparent.

How did it hold up? This car wasn't babied as a new Lincoln owner might break in his car. We took railroad crossings at speeds we approached them; we criss-crossed over rutted roads, gave the car little reprieve between acceleration runs. Yet nothing came undone; engine noise remained at its low level; chassis, transmission, and running gear did not complain. Slowing down in LOW gear from a ¼-mile run possibly ruptured a plate in the left-side muffler, for the tailpipe emitted a few soft explosions not evident earlier; but the normally quiet exhaust system never raised its voice during the test.

There was evidence of staining around the rear bumper exhaust outlets during the tests, but it did not reappear in normal driving conditions. A note of warning: Stay clear of Lincoln's high-set exhaust pipe outlets if you're walking thru a parking lot or working in a gas station. After high-speed running or the dynamometer check, these chromed outlets became dangerously hot—and we mean hot enough to burn our hand.

Servicing: Altho engine changes are among the '56's differences, these won't change servicing, for most components remain where they were in '55. Something that servicemen will find unusual is the carburetor air intake, a tube running from behind the radiator crossbrace up to the aircleaner. In this large hose is a choke-like heat valve controlling air intake to the carburetor. It admits warm air from an exhaust manifold heater when carburetor air temp is 65° F or below, and from cold air intake at 90 and above. This will undoubtedly give quicker and more satisfactory warm-up, prevent carburetor icing and its attendant stalling, and increase engine power due to cooler intake air. A 50° F reduction (not unreasonable) gives about 4 per cent more power.

SUMMING UP

Lincoln is prepped for a prominent spot in '56 fine-car sales picture. And it's not overlooking any of the current pitches; its safety program includes a "deep-dish" steering wheel (designed to keep the driver off the steering column in a collision), a non-glare, covered (not padded) dashboard, safety doorlatches, and optional seat belts. Its low, wide and handsome body drew many admiring glances (and close inspection) wherever we stopped during the test. Prices are competitive, and the big, quiet package offered is attractive in many different ways. We're inclined to predict that when sales are tallied at the end of the model year, you won't have to look too closely to recognize Lincoln in that gilt-framed picture.

—Jim Lodge

GENERAL SPECIFICATIONS
1956 Lincoln

ENGINE: Ohv V8. Bore 4.00 in. Stroke 3.66 in. Stroke/bore ratio 0.915:1. Compression ratio 9.0 to 1. Displacement 368 cu. in. Advertised bhp 285 @ 4600 rpm. Bhp per cu. in. 0.774. Piston travel @ max. bhp 2806 ft. per min. Max. bmep 164.7 psi. Max. torque 402 lbs.-ft. @ 3000 rpm.

DRIVE SYSTEM: STANDARD transmission is Turbo-Drive. 3-element torque converter with planetary gears. RATIOS: Drive 1.47 x converter ratio and torque converter only (2.40, 1.47 and torque converter only, at full throttle thru downshift detent); Low 2.40 x converter ratio; Reverse 2.00 x converter ratio. Maximum converter ratio at stall 2.1 @ 1550-1750 rpm.

REAR-AXLE RATIOS: Standard 3.07, 3.31 with air conditioning.

DIMENSIONS: Wheelbase 126 in. Tread 58.5 in. front, 60 in. rear. Wheelbase/tread ratio 2.13. Overall width 79.9 in. Overall length 222.8 in. Overall height (empty) 61.2 in. Turning diameter 45 ft. 7 in. Turns lock to lock 3⅓ (power steering). Test car weight 4675 lbs. Test car weight/bhp ratio 16.4:1. Tire size 8.00 x 15 (tubeless).

PRICES: (Suggested retail price at main factory; does not include federal tax, delivery and handling charges, or freight.) CAPRI 4-door sedan $3821, hardtop $3735. PREMIERE 4-door sedan $4183, hardtop $4183, convertible $4318.

ACCESSORIES: Turbo-Drive, power brakes, power steering standard, radio (with automatic antenna, rear-seat speaker) $126, heater $125, power seat only $74, power seat and power windows (standard on Premiere), $186, air conditioning $556.

ROAD TEST — **AUTO AGE**

Which is the better car—Lincoln or Mercury? Here's what we found out. . . .

PROVING GROUNDS REPORT:

LINCOLN and MERCURY

116

WHEN WE FIRST SAW the completely new Lincoln for 1956, several members of our staff remarked that they were curious to find out if it would bear more resemblance to the '55 Lincoln or the '56 Mercury in a road test. Perhaps this is a snide way of looking at things, but for years people have been calling Mercurys "bigger Fords" and Buicks "poor men's Cadillacs" and things of that sort, so we wanted to get the real scoop in order to be able to give a fair report, even to skeptics.

We were fortunate, therefore, in being able to test both the Lincoln and the Mercury "back-to-back" on the Ford proving grounds in Dearborn, Michigan. As it turns out, we actually got to drive the Mercury Montclair first, but since the Lincoln was the only completely new car, we would rather concentrate the earlier portion of this report on the Premiere that we put through seem almost to take the place of fenders; actually the two merge into one sweeping unit to dominate the entire front end of the car. Even the huge wrap-around windshield goes unnoticed at first. This is a fine tribute to Lincoln styling, by the way, because wraparound windshields have a way of "taking over" the lines of most cars. They often seem out of place. This is not true with the Lincoln.

The only other styling feature that really hits you is the extremely long tail. Since the Lincoln's wheelbase is only 126 inches, this is obviously where they stretched the car to get that impressive over-all length. Esthetically this is the one thing we don't like about the car, but even if you agree with us you'll soon forget about esthetics when you see that fantastic trunk. You could almost carry a spare car back there instead of a spare tire.

Big Lincoln Premiere had no trouble at all with hill on proving grounds graded at a rugged 30 per cent.

Test car was taken over a variety of artificial cement bumps, handled beautifully even at relatively high speeds. The riding qualities are fantastic.

its paces. (Don't worry Melvin, we'll get to the Merc!)

As far as performance and handling were concerned, Lincoln need not have made any major changes over their 1955 product to stay ahead of the pack. But appearance, it seems, is more than half the battle when sales are at stake and so the Lincoln styling department went to work with a vengeance. They had the word "prestige" foremost in their minds before they had put a line down on paper and it was no secret that they were out to "out-Cadillac" the Cadillac. So people wanted a big car, eh? All right, they'd have it. The Lincoln for '56 may not be the very biggest car on the road, but its 18½ feet of over-all length make it a pretty strong challenger. And if it's weight you want, it's got that, too, close to 5,000 pounds of it.

But while inches and pounds seem to go hand in hand with "class," an equally important factor is beauty of line or, to use a broader term, styling. The big Lincoln has sure got styling, but whether you like it or not is going to be largely a matter or personal taste. This car is totally different in looks from others in its competitive price class and while it is certainly sleek and modern, it manages to be both simple and extreme at the same time. Viewed from the front, the car has a "horizontal" look, the bumper, grille and headlight assembly all blending into a pattern of bold, uninterrupted sweeps across the car, like chrome layers. Even the parking lights are long and narrow—and horizontal.

From the side, the large, forward-leaning headlights

But enough about styling. How does the car perform? Well you may remember last year that we said the Lincoln was the best-handling big sedan we had driven —until the torsion-suspended Packard came along. This statement still holds true (so far) for 1956, in spite of the fact that this car really *shouldn't* handle as well as last year's much-smaller Lincoln. The only reason it does handle as well (but not better) is because certain minor changes were made in the front-end suspension to compensate for the added weight and difference in balance. The result is a car that actually understeers just a little—that is, it breaks away at the front wheels before the tail decides to come around—thus giving you lots of warning if you go into a curve too fast. The steering itself is of course extremely light and surprisingly positive for a power-assisted unit, but still a bit slow for any of the really high-speed driving you are tempted to do in a car with this much power.

We took the car around the Ford handling course again and again, going faster and faster each time, until the tires were setting up a constant scream, but even when we hit a curve too fast we are able to "put the car sideways" as it were, sort it out, and stay on the road without too much trouble. Again, all the power helped because you can use power to advantage in a curve in a car that understeers. All things being equal, the more power you apply the faster your steering wheel snaps back to a straight position. This can save quite a bit of elbow bending in an automobile that steers over five

117

Here is another one of the cement obstacles on the test grounds. The Lincoln didn't slide at all, finally bottomed here at around 50 mph.

For every gain there has got to be a loss. While the car does handle well, its soft springing caused it to lean considerably on turns.

turns from lock to lock (with or without power assist).

We got a real charge—in more ways than one—when we took the car out on the high-banked speed track. We had been doing some pretty severe brake tests just prior to this and the brakes were still quite hot. It wasn't until we really got some speed up that we realized just how much they had faded, and then we decided to do a couple of fast laps at about 80 mph without using the brakes at all, to cool them off. This we did, and at that speed on the soupbowl the car didn't feel as though it was going any faster than about 55 or 60 mph on a normal highway. But then, when we figured the brakes were back to normal, we decided to open the car up to see what it would do under a heavy foot. Coming out of one long and sweeping turn into the straightaway at the front of the proving grounds, our test driver floor-boarded it at about 95 mph. Instantly we began to accelerate and the speedometer was still climbing as we went through the wide radius, full-bore turn at the end of the straight. It was just creeping past the 115 mph mark when the driver started backing off for the sharper turn ahead. Or at least he tried backing off. The brakes had other ideas; they seemed to have lost at least half their power and we barely got down to a "safe" 95 mph to get through that corner. In all fairness, though, those brakes had been sorely overstressed under conditions that would almost never be encountered by a car on the highway. Besides that, our Lincoln was a proving grounds test car; there's no telling how hard it had been driven before we got it.

Back on straight and safe roads again, we ran through our acceleration runs. Here are the times we got—all with a corrected speedometer:

Zero to 30 mph took, on the average, 4.6 seconds. We got one run down as low as 4.2 and a few others ran as

Up on the high-speed banked track the Lincoln, with 285-hp engine, was at its best. After several tries, speed was finally timed at 115 mph.

Mercury Montclair, above, handled and cornered well, but redesigned shock absorbers made ride softer, consequently the car leaned more.

high as 4.9. To 60 mph took 12.4 seconds with runs from 12.1 to 14.2, depending on which combination of low and drive we used. We found it best to shift out of low at about 48 mph, just before the engine peaked. Our 50 to 70 time worked out to about 6.5 seconds.

If you think these times aren't fast enough, take a look at some playing around we did with the Mercury and you'll learn how easy it is to have a variation in acceleration times of more than a second in either direction. There is more than one way to make a car—or a driver—seem good or bad.

In any event, speed champion or not, we found the 1956 Lincoln Premiere to be a lot of automobile from any standpoint—a real pleasure to drive. It's well made inside and out and the upholstery and paint color options are almost endless. Seating position for the driver is quite excellent and visibility is all that could be asked for. Comfort and a luxury feel are this car's strongest points, the ride being awesomely good, even over the cobblestones and other bad-surfaced test roads on the Ford track. Perhaps most impressive of all was the way the car could be driven at relatively high speeds over these bad roads without any wandering, sliding or pitching. The 1956 Lincoln seems ready, willing and able to perform any highway task. How do we rate it against the Mercury Montclair? Let's be fair and tell you about the Mercury first.

THERE IS little that can be said about the appearance of the 1956 Mercury Montclair. This is not because we do not like the car; on the contrary, we voted the 1955 Montclair the best-looking American automobile of the year. Since it hasn't really changed in appearance for '56 it's pretty obvious that we approve of the styling. The only thing that's really been done on the outside is to lower the roof line on sedan models ¾ of an inch, but you'd never notice that. Inside, you're sure to notice the new "safety-dished" steering wheel.

The first thing we did with the Montclair was to take it out on a grand tour of the entire Ford test grounds, to see how it felt on all types of road surfaces, banked and unbanked curves of varied radii, and grades. Around the flat, looping handling course the car felt a bit softer than it had last year, and when we checked with a Mercury engineer we learned that the shock absorbers have been reworked slightly to give the car a softer ride. This has been done to appeal to the upper-middle-class-type automobile buyer who has been demanding more and more luxury each year, but we don't see any good reason for going backwards in suspension design. If the engineers could have found a way to make the ride softer while keeping last year's handling qualities—or even improving on them—we would have been in accord. But they know what they are doing, really; most people who buy Mercs want the biggest, plushiest ones they can get.

After we came off the handling course—where the car actually behaved quite well, displaying a pleasant amount of understeer and correction control—we headed for the big hill that is graded at 15 per cent on one side and a whopping 30 per cent on the other. The car shot right up the lesser grade with very little head start so we backed down, drove up again and stopped in the middle. There the car started from a dead start

Braking tests were run off with these white lines as guides. Power brake for '56 has valve on the air reservoir, holds power over night.

as though it didn't know there was a hill at all and we were suddenly anxious to see how it would perform on the 30-per-cent grade.

We came at this frightful monster at about 30 mph in drive range and started up at constant speed. As the grade stiffened we fed more gas until the accelerator was right down to a point just above the kick-down gear; in other words it was getting lots of gas but still in high. It started to lose speed and continued to do so until the half-way mark when our driver suddenly mashed his foot all the way down. The Mercury dropped into a lower gear and accelerated right up over the crest of the hill. We drove down the other side and came around again, coming part way up and stopping half way up. When the throttle was floorboarded the Merc took off so fast that there was actually some wheelspin—this on a 30 per cent grade!

Thus satisfied with the car's hill-climbing abilities, we drove out onto the speed track. Here the car performed more or less the same as last year, only faster. There was no wandering at all on the straights and although the steering did have that "lost" feel common to all power-assisted units at high speed, there wasn't much of a cross wind and so correction wasn't necessary. We tried the brakes several times for fairly high speeds and they seemed quite good; no grabbing, fading

were the zero to 60 times. We got to an actual 60 mph in 12.2 seconds but to an indicated 60 in just 10.1 seconds. Our honest 30 to 50 time was 5.4 seconds.

But there's more. Even with a corrected speedometer you can get quite a variation in acceleration figures just by punching the stopwatch at different times. A way to get "fast" time is to wait until the car actually starts to move before hitting the watch. Taking into account driver reaction time and mechanical lag (especially with an automatic transmission) you may gain a second or even a fraction more by this method.

But to be fair, we used the corrected speed readings, starting the car and the stopwatch on exactly the same signal. Please don't write to us complaining that our times for the Mercury were "slow" (or for any car we happen to test) because if you want to fool yourself into thinking any given car will go faster than it really will, who are we to stop you? Before you get indignant, however, please try to remember that our acceleration and speed figures are usually within a few tenths of a second of those gotten by the test engineers themselves on the company tracks.

At the conclusion of all the various speed tests, we went back to the bad-roads section of the proving grounds. Admittedly, most of these surfaces are worse than those you would subject your car to, even if you

Mercury styling for '56 is basically unchanged from last year. Grille and chrome trim have been reworked, sedan roof lines lowered a bit.

Taken over the railroad tracks at speed, the car threw passengers up toward roof but did not bottom. Safety belts did their job well here.

or sharp nose-diving was experienced. As to the Merc's top speed, we got a good 108 out of it. You could probably do even better on a long stretch of level road—say up to 110 or so.

With the speed runs completed and the car nicely warmed up and running like a clock we swung into our acceleration tests. First thing to do was to check out the speedometer, and perhaps if we give you the figures we got you will begin to understand why there can be so much variation in road test times, or in claimed highway speeds. When the speedometer read 10 mph, actual speed was 9.7; at 20 it was 18.4; at 30 actual speed was 27.8; at 60 it was only 52.8; when it read 80 we were really going 71.4 mph and at 90 the true speed was 81.6. And this is the sort of error you are liable to encounter with any car.

What did the variation do to our acceleration times? Zero to an actual 30 mph took 4.7 seconds but when we ran the same test going by indicated speedometer readings the time was 4.3 seconds. Much more shocking

should happen to run into any similar ones. The average Joe, with a car he planned to keep for awhile, would simply take another route. Ordinarily we would, too, but the tests were a lot of fun, and, more important, they provided us with a better idea of just how far these cars can be pushed before they will balk. We went over some cross-mounted railroad tracks at about 35 mph and would have banged our heads smartly on the roof of the car if we hadn't been restrained by safety belts; but even at that the car didn't bottom. The same thing happened with a number of artificial cement bumps in the road. We had to go over a sharp, two-foot bump at 40 mph to finally make the Mercury bottom. The only difficulty we had at all with any of the bumps was keeping control of the steering as we bounced heavily on the front wheels during "landings." Under normal circumstances, the Mercury is one of the most manageable cars on the road today.

When we sat down to compare the Montclair and the Premiere, we had a tougher time than we had anticipated. We had picked the largest, most powerful car in each line purposely, so that we could make a proper evaluation of everything that the Lincoln and Mercury had to offer. What we discovered was this:

The Lincoln is an entirely new and individual car. It does share certain mechanical features with Mercury, last year's Lincoln—or even Ford, for that matter—like ball-joint suspension and a torque converter transmission with planetary gears. But this does not mean that the Lincoln is a high-priced Mercury any more than you would rate a whale a high-class shark just because it's bigger. The point is, each car is aimed at a specific class; they are designed for different purposes. This was not only the intention of the manufacturers—it actually worked out that way. It might *not* have if the designs weren't so carefully executed in both cases. A bigger car is not always a better car, and the Lincoln is not really better than the Merc unless it suits your purposes better. Both are beautifully constructed automobiles that will give years of trouble-free service to any owner who takes the trouble to care for them properly. If you want a smaller, sportier car with lots of performance and a feeling of class as well, the Mercury is your dish. If you want a real dreamboat that will look well with your yacht and house in the country—and yet will perform with the best of them under the toughest of conditions, then the Lincoln is probably what you have been looking for. No matter which you choose, you can't go far wrong.

We took a vote among our staff members to try and find out in a preliminary way which of the two cars *we* would choose for ourselves if pinned down. This experiment, sad to say, did not help at all. There is an *even* number of AUTO AGE staff members at present and they were split right down the middle in their choice. If we should happen to hire someone else in the next month or so, or fire someone—to make the number *odd*—we will then have a clear-cut victory for the Lincoln or Mercury and we promise to let you know about it right away. ●

After half an hour of severe braking the Mercury was still making safe stops, but some fade was in evidence. Note lack of nose diving.

LINCOLN PREMIERE SPECIFICATIONS

ENGINE: V-8, overhead valves; bore, 4 in.; stroke, 3.66 in.; total displacement, 368 cu. in.; developed hp, 285 at 4,600 rpm; maximum torque, 401 ft./lbs. at 2,800 rpm; compression ratio, 9 to 1; four-throat carburetor; ignition, 12 volt.
TRANSMISSION: Turbo-drive torque converter with three planetary gears.
REAR AXLE RATIO: 3.07 standard; 3.31 with air-conditioning.
SUSPENSION: front, independent ball-joint coil-spring system; rear, longitudinal, semi-eliptical leaf springs.
BRAKES: four-wheel hydraulic, internal expanding; 12 in. diameter drums; power booster.
DIMENSIONS: Wheelbase, 126 in.; front tread, 58.5 in.; rear tread, 60 in.; width, 79.9 in.; height (loaded) 60.2 in.; over-all length, 222.8 in.; turning circle, 46.2 ft.; dry weight, 4,362 lbs.; tires, tubeless, 8.00 x 15.

PERFORMANCE

ACCELERATION:
 Zero to 30 mph: 4.6 seconds
 Zero to 60 mph: 12.4 seconds
 50 to 70 mph: 6.5 seconds
TOP SPEED: 114-115 mph.

MERCURY MONTCLAIR SPECIFICATIONS

ENGINE: V-8, overhead valves; bore 3.80 in.; stroke 3.44 in.; total displacement, 312 cu. in.; developed hp, 225 at 4,600 rpm; maximum torque, 324 ft./lbs. at 2,600 rpm; compression ratio, 9 to 1 (8.4 to 1, optional); four-throat carburetor; ignition, 12 volt.
TRANSMISSION: Mercomatic torque converter with three planetary gears; standard three-speed manual and overdrive available.
REAR AXLE RATIO: conventional, 3.73; overdrive, 4.09; Mercomatic, 3.15.
SUSPENSION: front, independent ball-joint coil spring system; rear, longitudinal, semi-eliptical leaf springs.
BRAKES: four-wheel hydraulic, internal expanding; 190.90 sq. in. braking area.
DIMENSIONS: wheelbase, 119 in.; front tread, 58 in.; rear tread, 59 in.; width, 76.4 in.; height, 58.7 in.; over-all length, 206.4 in.; turning circle, 43.19 ft.; weight, (dry) 3,541 lbs.; tires, tubeless (extra low pressure), 7.10 x 15.

PERFORMANCE

ACCELERATION:
 Zero to 30 mph: 4.7 seconds
 Zero to 60 mph: 12.2 seconds
 30 to 50 mph: 5.4 seconds
TOP SPEED: 106-107 mph.

Continental on a moderate corner shows slight body lean to the camera. Its handling and ride are unusual for a car of its size.

Driver's Report
THE NEW CONTINENTAL MARK II

$10,000 is a lot of money for an automobile. What do you get in performance, roadability and quality? Here are the answers

BY GEORGE KNIGHT

Powered ventilator interferes with placing arm on window sill. Note novel handle and dirt-catching cups for door grips.

Side mirror bracket is formed as part of windshield molding. Doors, and all body joints, have an exceptional precision fit.

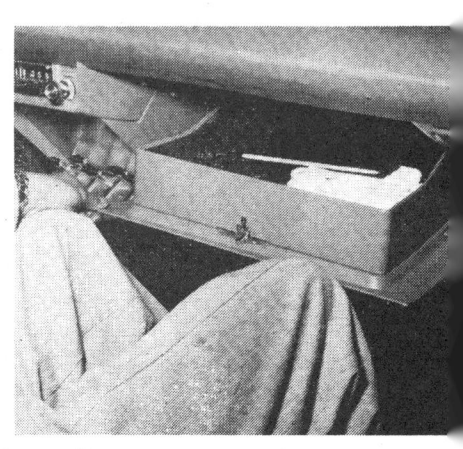

Righthanded glove compartment drops out of dash, comes close to passenger's knees. Note hard-cover owner's manual.

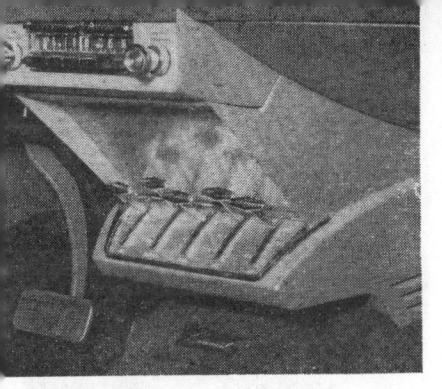

Glimpse of things to come as Continental's controls move to pedestal-type hump—a feature on many dream cars.

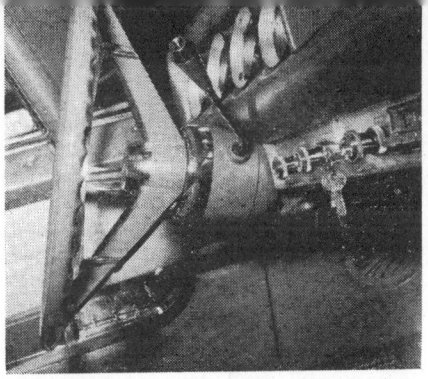

Driver feels closer to car with 17-inch wheel set up to dash. Small instruments have faces like fine, jewelled watches.

Single lever controls both latch and safety catch on the hood. The metalwork, in general, is above average of U.S. cars.

LATE last fall, after years of speculation and tantalizing rumor, the new Continental Mark II finally appeared. Since its price tag of $10,000 makes it unattainable for all but the very well-heeled, most of us have had to be content with merely looking at the car. Even obtaining one for test purposes was blocked: the factory declined to make one available on the grounds that a low production rate (about 2,000 have been built to date) was geared to orders; and the rare owner one might find was understandably reluctant to submit his prize to strenuous road testing.

Recently, however, George Barris, of customizing fame, decided to invest in a Continental—with professional motives in mind. Less than 10 minutes after he had received the car from an authorized dealer, who had driven it for 7,000 miles, he agreed to make it available to MOTOR LIFE for study.

The Continental has a factory-recommended price of $9,941; with air conditioning, the only optional extra, it is $10,681. This puts it in a class by itself among U.S. cars, the next costliest production model, an eight-passenger limousine, running more than $2,000 less. For this kind of money a buyer expects to get something pretty special.

Performance has been one of the biggest question marks behind the Continental's name. The builders do not quote horsepower or torque output, but it is estimated that the yield is not far above that of the 1956 Lincoln engine, which the car uses, with minor modifications. So with somewhere in the neighborhood of 300 hp underfoot, we took the Con-
(Continued on next page)

Examples of unity of styling on the Continental are front parking lights and rear exhaust outlets. Bumpers also are identical, an extremely rare feature on U.S. cars.

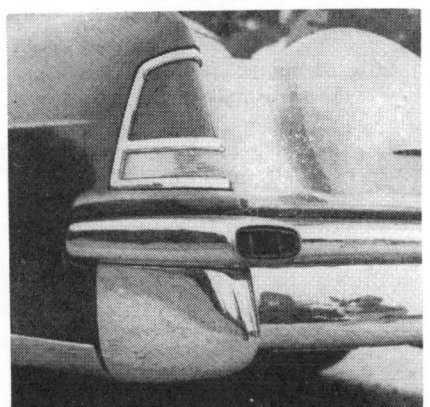

Gas intake is concealed behind the unusual hinged left tail light assembly.

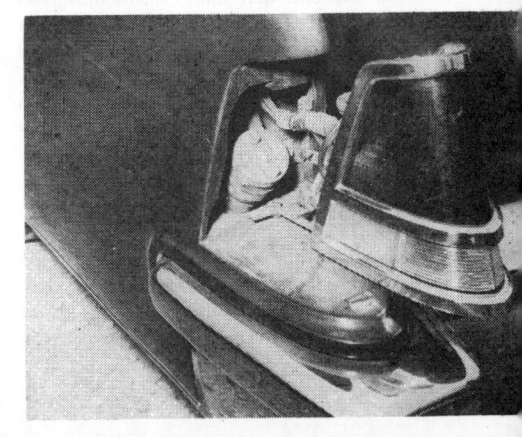

George Barris admires the headlight on his Continental. The unit is conservatively recessed, rather than being radically Frenched in the current fashion, but up-to-date.

Wheel discs are not a close fit, cause many to wonder if they are coming off. Continental designers considered genuine wire wheels, but finally made the concession to costs.

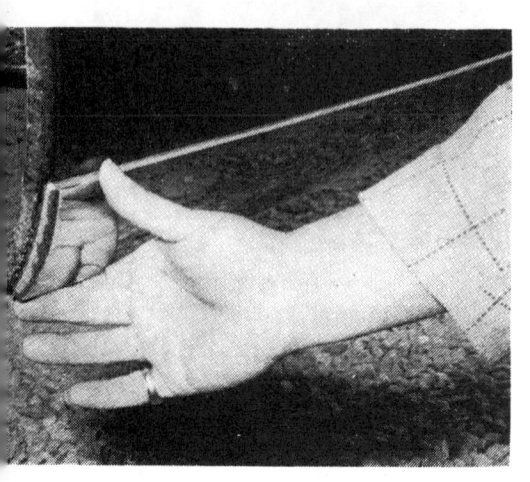

Chrome strip has not been bolted on haphazardly; body metal is recessed.

tinental out of the road and put it through standard road test acceleration techniques which involve various shift combinations.

The quickest times were: 0-30 mph in 3.9 seconds; 0-45 mph in 6.9; and 0-60 mph in 10.5. In the latter category the gear lever was held in low all the way and the tach was registering 4900 rpm at the top end. In an alternative check, starting in drive and allowing the torque converter to shift midway, produced a time of 11.5 seconds. From this it is evident that the Continental was devised to perform adequately, but not spectacularly. The 0-60 mph elapsed time is a good two seconds slower than the average hot standard passenger job and about on par with most of the big V-8 machinery. Weight, of course, penalizes the Continental, since it will run from 200 to 400 pounds over the 5,000 mark in normal trim.

In handling and ride, however, the Continental outclasses any of the non-specialized production cars. The power steering has snap and precision, with excellent feel of the road. Going through a series of winding curves at 80 mph is deceptively easy, with behavior similar to that of any good passenger-bodied car at, say, 40 or 50 mph. The security and road-hugging qualities are superb. Only on a near right-angle bend at abnormal speed does the mass of weight thrust out over the wheels and fight directional change. Even then, however, the heeling sensation is slight.

Contributing to the feeling of maneuverability is the positioning of the driver close to the dash, creating the illusion of handling a much smaller car. The steering wheel nearer the vertical adds to the effect. The closeness of the roof and side pillars also provide a compact arrangement. On the other hand, with the floor inside the frame rails, the Continental is not the easiest car to get in and out of. As a sidelight, it was noted that with windows rolled down, the level of wind noise was high, although when completely closed the car was exquisitely smooth and quiet.

What a buyer gets in performance and roadability, however, is only a fraction of the story. The styling is conservatively modern, but does not offer anything that is radically new. The big point with the Continental is its utterly simple and functional layout that employs materials of apparently exceptional quality. As many of these details as possible are shown photographically with this text. But the richness and attention to detail must be seen to be appreciated. It amounts to a new standard of finish for U.S. cars and this, along with the outstanding ride and steering, is what the buyer gets for his money. Is it all worth $10,000? Probably not, since such items should be available at a much lower price (and are in some inexpensive foreign cars). The only thing left then is the prestige of driving such an exclusive car. But this is where tangible values end and salesmanship begins. •

Floor-frame layout caused exhaust lines to be run outside rails which results in some warming-up of the rocker panel and covering chrome strip on the exterior.

Fuses are most conveniently located on engine side of firewall, with box cover identifying each. Note empty brackets, probably built in for future accessories.

Fitting 5th wheel to car was problem, since rear bumper fits to body without normal gap. Metal meets metal in lip.

Spare in trunk leaves minimum of luggage room. Compartment is so air-tight that it could use moisture ventilation.

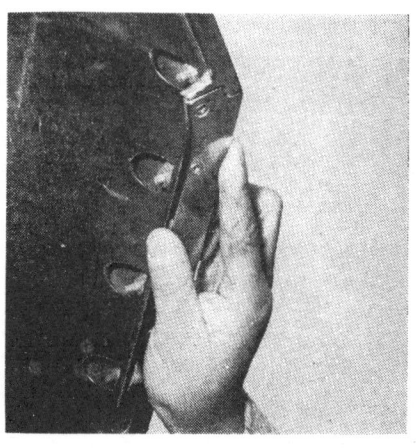

Another example of fine attention to detail: hood edge is heavily chromed strip showing smooth joint and flush screws.

Cruising at speed in the Continental is unsurpassed for effortless smoothness, 100 mph being no more strain that half that much in the average car. It is best example of "like riding on rails." With windows open, however, wind noise reaches high level.

Adjustment of A-frame is from inside engine compartment. Jack, also, is novel, is operated by crank at waist height.

Heavy mesh liner, instead of usual padding, soundproofs Continental's hood. Metal strips for fit are an extra touch.

View of frame shows how floor setting down between members requires new routing of exhaust, linkages, extra U-joints.

Unusual mechanical feature is right-angle frame member in front especially designed for low-silhouette requirements.

Unconventional exhaust system places mufflers inside fenders behind rear wheels, with additional small resonator (not visible) in front of wheel, and crossover pipe to other side.

Like Lincoln, of course, the Continental engine has the new thermostatically-controlled intakes which help regulate temperature of air to the chambers. Large hose inducts cool air.

'57 LINCOLN

AN MT RESEARCH REPORT
by John Booth

William E. Burnett
Executive Engineer
Ford Car Engineering
Ford Division

MT DETROIT INTERVIEW:

NO ASSIGNMENT is more calculated to please a *working* engineer than to be told to build a car that is new from tire tread to roof. For an executive engineer in charge of Ford passenger car design like big Bill Burnett, this opportunity has attained a frequency of once every two years. It means 16-hour days and aspirin by the gross, but he thrives on the routine.

Burnett is one of those rare engineers who are completely articulate. He is a popular and sought-after speaker on the SAE circuit, and the captive audiences at these affairs brighten visibly when he ascends to the lectern. Although now in charge of all phases of Ford car engineering, Burnett is primarily a chassis and ride specialist.

He received his early training with Chevrolet's Ed Cole (see page 12) at Cadillac under Owen D. Nacker (retired). Nacker, nicknamed the Sheriff because of his affinity for a disreputable Stetson which he reputedly wore even in the bathtub, was famous for his ability at ride evaluation, a seat-of-the-pants procedure.

Bill Burnett is perhaps most pleased that he was able to engineer the new car on two wheelbases. While the decision stemmed primarily from sales (to give Ford a broader coverage of the market despite the forthcoming "E" car), he engineered into the longer of the two a ride which he will stack against anything being built today. What he calls the "pendulum effect" (a system of forces produced by the relationship of body overhang to the wheelbase) is closer to the ideal than ever achieved in past Fords.

Despite a softer ride, Burnett claims to have maintained all that was good about Ford handling, and in the past, this has been good enough to win a number of MT awards. Unlike his contemporaries in other Ford divisions, he favors building in a slight amount of oversteer, that characteristic which will cause the rear end to break loose before the car itself during hard cornering. This, coupled with a reasonably soft ride, he says "is the type of control best adapted to universal use."

The recent pilgrimage to Detroit of a group of safety-minded congressmen has scared everybody concerned with putting more power under the hood, and Burnett is no exception. Nevertheless, the company and its divisions have no intention of keeping mum about fast-selling horsepower. Ford offers the widest spread of engines of any single brand of car. There is a six, and three varieties of eights, with every transmission option on each. And there's the racing package, undoubtedly to be introduced just prior to Daytona.

—Don MacDonald

—TOP-DRIVING U.S. CAR?

PHOTOS BY JOE FARKAS

WEBSTER defines the word *refinement* as "affectation of elegant or subtle improvements." We define the '57 Lincoln as being that which typifies refinement. From the "Quadra-lite" front end to the canted rear fender blades, the new Lincoln is chock-full of styling and engineering niceties to further enhance its foothold on the fine-car market.

Style-wise the '57 Lincoln has a new crispness to its lines while retaining the distinctive design theme that made the '56 so impressive. Engineering-wise, mechanical improvements are highlighted by more docile responsiveness and improved ride characteristics—all lending to the feeling of fine car luxury.

Interior appointments remain much the same as last year. Instruments, power window controls, wide "dual-lever" brake pedal and shift control are all reminiscent of the '56 with subtle color or trim changes that create an overall measure of "newness."

Lincoln's success with its mammoth sweep-hand speedometer has dictated a repeat in '57. Like last year, functional simplicity of the instrument panel lends an air of planned elegance that is carried out in complementary appointments throughout the interior. Innovations include repositioned inside door locks on the panel side of the window sills, which operate forward and backward, and doorhandles that slant outward and are large enough to really grab hold of. Upholstering materials (including top grain leather) are offered in many colors and textures creating a certain amount of individualism in interior selection.

The Lincoln Premiere series has as standard equipment, power seats, windows and brakes (all available as extra equipment on the Capri). New powered extras available this year operate the vent windows, six-way power seat and door locks, the latter operated from a master switch on the instrument panel.

Other extras for '57 include Adjust-O-Matic shock absorbers and a power-directed differential which prevents the car from being immobilized when one wheel is stuck in mud or snow.

A remote control side mirror and an automatic low-fuel warning signal that flashes red when three gallons of fuel remain are standard equipment, as are the deep-dish safety wheel and safety door locks. Belts, padded instrument panel and sun visors will complete Lincoln's safety package but come under the heading of extra equipment.

Performance figures won't show any appreciable difference over the '56. Our acceleration test was made with a fifth wheel and a bank of stopwatches, making the procedure as accurate as our full-scale tests. Our best times (with or without manual downshift) were as follows: 0 to 30 mph in 4.3 seconds (about 0.2 less than last year) and to 60 mph in 11.5 seconds (same as '56).

At cruising speeds, acceleration is again in the '56 class with four seconds required to get from 30 to 50 and 12.2 seconds needed to get from 50 to 80 mph.

continued

'57 LINCOLN continued

Rear quarter view is still unmistakably Lincoln. The '57, left, doesn't appear very different until you view the canted fenders from the rear.

The '57 Lincoln engine remains basically the same as last year. Horsepower has been upped from 285 to 300 at 4800 rpm and torque has been increased to 415 pounds-feet (401 last year). Cubic inch displacement remains at 368. This added horsepower was accomplished by better breathing, a new distributor and a compression increase to 10.0 to 1 (was 9.0 to 1) through slight piston and combustion chamber modification.

Refinement rather than complete newness is apparent in many engine and accessory modifications which, while not necessarily of major importance, do contribute to smoother trouble-free operation.

A new carburetor designed to withstand heat-soak conditions without boiling dry reduces stalling and hard starting due to heat (vapor lock). Fuel baffles or dams located in the primary jet area of the carburetor main body have reduced flooding or starving of the engine in high speed turns, a universal complaint of four-barrel carburetors (See Sept. MT "Technical Questions"). A new type of "Paper-Pak" aircleaner reposes atop this carburetor and is said to be 99.5 per cent efficient as compared to 98 per cent for the conventional oil bath type. The cellular core can be cleaned by gently tapping it on a solid surface, or easily replaced with a new element which is never oiled.

Another improved item is a new full-flow throwaway oil filter. Composed of five parts sealed in one shell, it features "light bulb ease" of installation and removal. This can even be accomplished without undue contortion as it is readily accessible.

New three-piece oil control rings, self-locking tappet adjustment screws and a smaller (12-inch) steel torque converter with front-mounted oil cooler complete the engine refinement features.

Roadability in the '57 Lincoln has been given a shot in the arm. Not that the '56 had to take a back seat to anyone in this department, but it's still better this year. The ride remains soft but not mushy, and oscillation is conspicuous by its absence even over coordinated bumps that normally make a car act more like a bronco. We whipped this new Lincoln into really tight turns at spine-tingling speeds, and the only protest to be heard came from a set of screeching tires. Definite understeer was noted along with minimum body lean; the car goes into a four-wheel drift rather than just breaking loose at the rear end. When we

The new four-door hardtop series has the same commodious proportions as its conventional centerpost counterpart.

128

For better efficiency in the new 12-inch steel torque converter, '57 Lincolns have an air-cooled oil radiator.

purposely ran afoul of the basic laws of physics, the resultant skid was easily checked and directional control regained by a quick twist of the power-operated wheel. There appears to be a deliberate attempt to compromise between oversteer and understeer.

We braked and violently accelerated in several turns and noted the inherent stability that has been accomplished by careful "tuning" of the chassis. The whole car gives the sensation of being really tied together with all components content to go in the same direction at once. A most satisfying phenomenon!

Road feel has been preserved for those of us who like to think a car should be steered, not aimed, and road shock is effectively dampened out. This power assist has enough oomph to make parking easy, but at the same time you will be aware of a certain wheel resistance in a static turn.

Directional control is not affected unduly by cross-winds (we had 25-mph winds for a while) but the correction in varying

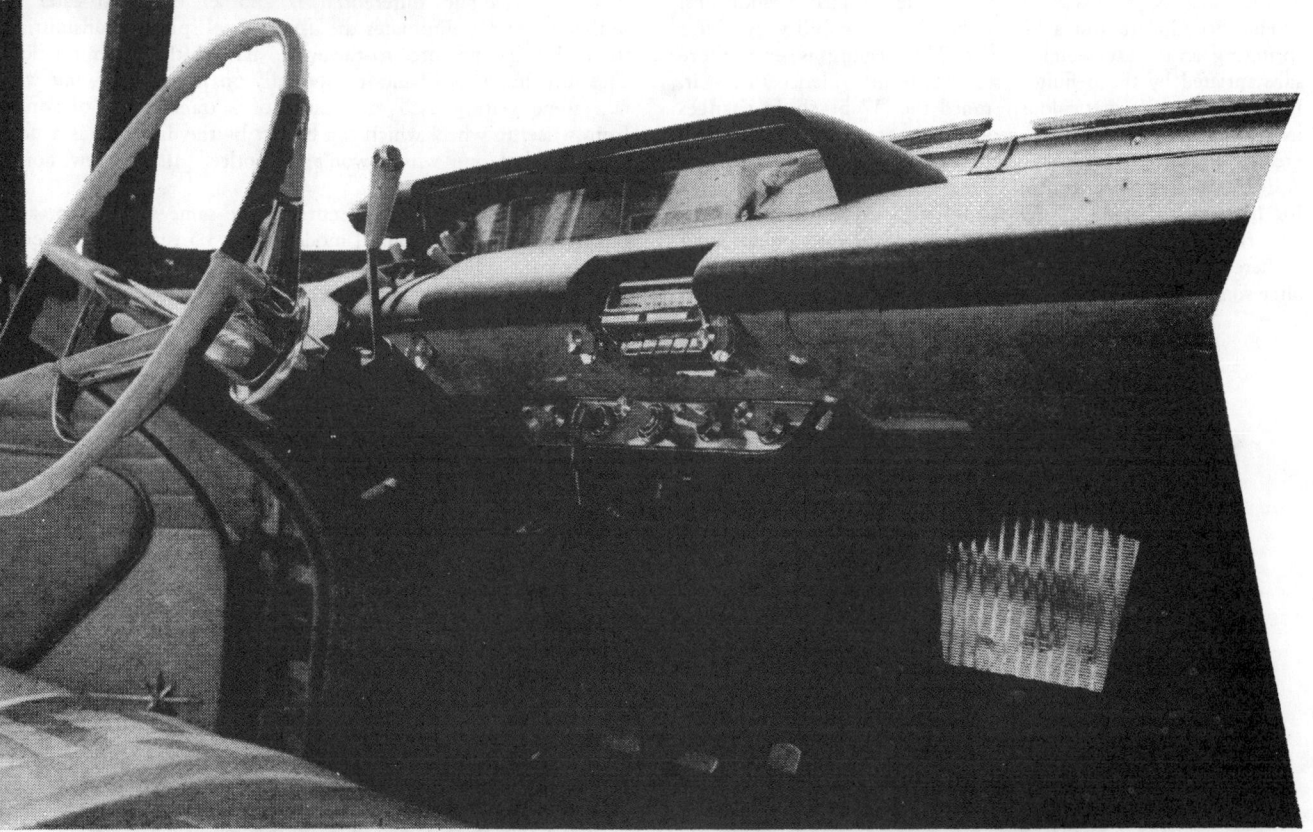

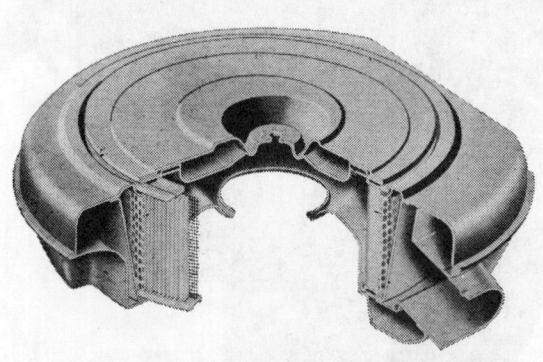

Almost impossible problems with the aircleaner arose as a consequence of the '57 Lincoln's low hood lines. A solution was found in the use of a simple and efficient new Paper-Pak dry type unit shown in the cut-away view above and installed at the right.

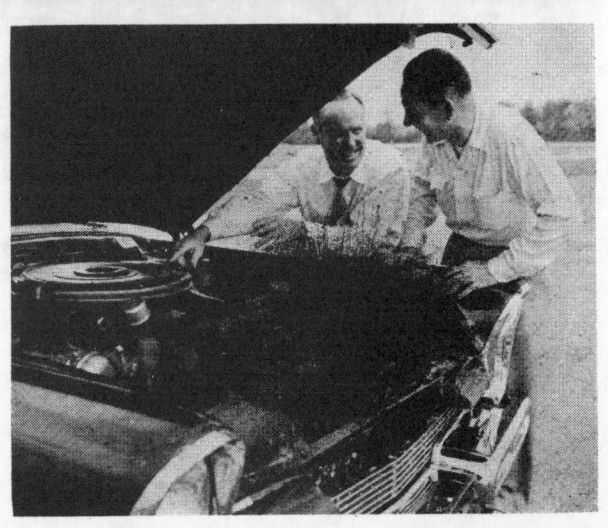

129

'57 LINCOLN continued

The new '57 Lincoln Premiere convertible.

degrees will be necessary, probably due to gusty conditions.

The '56 Lincoln lost a bit of the '55's roadability by softer springing and body weight shifts. This feeling is even more substantiated by the definitely softer ride in evidence this year. Lincoln engineers meticulously tuned the '57 by slight changes in spring rate and steering geometry, then tied it down with greatly improved Hydro-cushion shock absorbers. They are designed to give better control over rebound and can be adjusted for hard or soft ride at the factory or local Lincoln dealer's. Unfortunately, this variable control isn't on the instrument panel (where it should be) but the adjustment is simple and does offer some choice in ride characteristics.

The limited-slip differential is another effort to cater to individualism. It eliminates the usual wheel spin by transmitting the major driving force to the wheel having the better traction. This unit has a lot of merit, especially on ice or snow, but will take some getting used to because of a transference of forces from wheel to wheel which can be felt by the driver. It is a new sensation but one which won't be noticed after a few hours of living with it.

Power steering remains essentially the same as last year and is standard equipment on all models with 3.5 turns lock to lock.

The frame under the new Landau four-door hardtop (available in both the Premiere and Capri series) has been substantially

A smaller, more efficient steel torque converter transfers power from the 300-hp engine to the rear wheels in the smooth, quiet manner expected in a luxury automobile.

130

beefed up to retain body rigidity. The regular four-door has a very thin centerpost with window frames designed in such a way that you won't find it easy to distinguish the conventional sedan from the Landau with the windows up.

The fad of exhaust outlets in the rear bumpers appears to be on the way out. Lincoln, like other makes, has concealed them under the car this year, and we feel this is a step in the right direction. At least the unsightly discoloration around bumper outlets will be eliminated.

Overall body length has been increased to 224.6 inches but height remains the same as last year (60.2). Lincoln has retained the 15-inch wheels which carry 8.00 tires (8.20 on convertible and air-conditioned models).

All in all, while the new Lincoln can't be called completely new, there are enough changes both style-wise and mechanically to create interest. We feel the canted rear fenders especially will come in for their share of pros and cons. Those who particularly liked the looks of the '56 will probably feel this new fender treatment has detracted from the quiet dignity that characterized its predecessor. Others, of course, will say it adds character. The same can be said for the front end treatment with the Quadra-lite theme. Four headlights are something you have to get used to and while we think they are here to stay, plenty of people will take the attitude that they look awkward. Aside from their looks, however, they have many safety advantages and should be looked upon as a boon to highway driving at night.

Regardless of individual likes and dislikes, the '57 Lincoln need not take a back seat to any car as one of America's top luxury automobiles. It is sleek, smooth and docile. It has a definite place in the hall of distinction. —J.B.

The new limited-slip differential (optional equipment) will greatly improve traction on slippery roads. When equipped with suburban or snow tires, this unit would be a definite winter asset in northern areas.

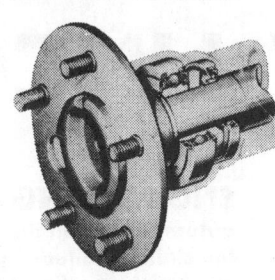

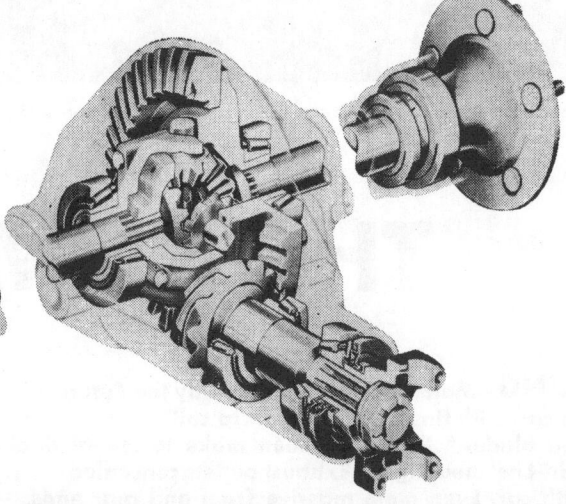

Modifications other than the outwardly canted fins include relocation of the exhaust tips under the bumper, directed downward, and placement of the back-up lights in the bumper.

A fake airscoop is used at the front of the rear fender fins.

Premiere two-door hardtop. Capri differs in hubcaps, lettering.

Characteristic long Lincoln lines have been broken for 1957 by lift of the rakish fins. This car is the new Landau hardtop.

THE NEW LINCOLN

STYLING—Major facelift influenced by the Futura dream car, with fins Lincoln prefers to call "canted blades." First production make to reveal dual "Quadri-Lite" headlights. Exhaust outlets concealed beneath car. Even more massive front and rear ends.

PERFORMANCE—More power enables quicker acceleration—about average for faster big-car bracket with 0-60 mph close to 10 seconds. Handling still better than its class.

ENGINEERING—The big pitch will be "the most automatic car on the road" and new gadgets support the claim. Limited-slip differential, plus automatic transmission refinements. Lincoln bypasses 14-inch tires.

BODY TYPES—Newest and most important is the addition of a Landau (four-door hardtop) to the line. Other types in the Premiere and Capri series are same as in 1956.

THAT the day of the minor grille-and-trim facelift is over is obvious when you compare the 1956 and 1957 Lincolns. The 1956 was new from the ground up. It was widely regarded as one of the most attractively styled cars of the year and set all kinds of new sales records for Lincoln. If you were going to pick a car on which to stand pat for another year, the Lincoln would have been a logical choice. Lincoln planners, however, decided otherwise.

The new Lincoln obviously could not be revamped completely after the major changeover a year ago. One glance will show that it has been changed considerably, however.

Most noticeable changes center around the rear end treatment—especially the rear fender or quarter panel lines. Anyone who has seen the Futura dream car will know where Lincoln stylists got their inspiration for the 1957 fender design.

The new fender line is higher than last year and doesn't flow straight back vertically as in '56. Instead, stylists have

Viewed from direct rear, fins have outward slant. Note novel nameplate mounted in "grille," absence of exhaust outlets.

Dual headlights have been widely forecast and Lincoln is the first into production with them. At right, Landau has an odd fin-like chrome panel mounted on the rear door. The air-scoop in the door is a non-functional item simply for looks.

introduced what they call a "canted blade" effect (they prefer that term to "fin"). The 1957 Lincoln rear quarter panels sweep up and out, making the car look wider and more massive when viewed from the rear. Like the Futura, they have an air scoop (simulated) built into the leading edge.

This new fender design is the most striking change from 1956—and will likely be the most controversial. Our first reaction was that it doesn't do as much for the car as the 1956 treatment. The '56 Lincoln was a distinctive looking automobile with clean, non-gimmicked lines. It looked long, low and big—but it was a graceful bigness. Some of this distinction seems to have been lost in the restyling. This is just the initial reaction, of course, and it may be that the new Lincoln look just requires some getting used to. Also, there will probably be many who prefer the canted blades to the more conservative look of 1956. Time—and sales—will tell.

Other rear end changes include the extension of the textured grille across the full width of the car. It also wraps around the sides of the fenders slightly. This change was made possible by the new bumper design which eliminates the exhaust pods. (Exhaust outlets are now concealed under the car.) The new bumpers incorporate wide, flattened-oval back-up lights at their extreme ends. Pyramidal tail lights are bigger and conform to the canted angle of the fenders.

Moving to the front of the new Lincoln brings us to another major styling change for '57—the dual headlight arrangement. (Lincoln calls them "Quadri-Lites.") Note that the lamps are mounted vertically rather than horizontally, as on most dream cars which have featured this treatment.

It should also be pointed out that these are not true dual headlights in the exact sense of the term. True dual systems use a pair of five-inch lamps on each side; one a long-range light, the other a short-range beam. In the Lincoln setup, the top lamp is a standard seven-inch unit with high/low beams; the smaller lamp below is an auxiliary driving light, more or less. This arrangement is used because obsolete laws prohibit a true dual system in many states—even though it would be much more efficient than present lighting methods.

Independent control systems are used for upper and lower lamps on the Lincoln. The seven-inch lights are operated by a normal instrument panel and foot dipper switch. The smaller beams are controlled by a separate switch on the side of the steering column. The two sets of lights are wired independently, but turning either of them on also turns on the tail lights.

How functional are the smaller, extra front lights? Well, they will be fine for town driving at dusk and at other times when you need a little illumination to increase your own visibility and to warn other drivers of your approach. They may be helpful on the highway under certain conditions also. Probably it was styling considerations and the desire to be ready when true dual systems are permissible that dictated the change, however.

Something else new is that all four-door Lincolns will look like hardtops this year—with the windows up, at least. Center pillars on conventional four-doors are quite thin and are concealed by the window frames when the windows are rolled up. This feature will appear on a lot of other makes and is a transition step toward complete elimination of all center pillars in the future.

News in a negative sense is that Lincoln will offer neither 14-inch tires and wheels nor push-button transmissions for 1957. (The 14-inchers apparently won't show up on many high-priced luxury cars this year.)

Exterior trim has been changed on Lincolns for '57—as was inevitable—but still has been held to a tasteful minimum. And aside from the dual headlights, there have been no particularly noticeable grille or front end styling changes. Overall height is still the same, just a fraction over five feet, but overall length has increased nearly two inches and width about an inch.

There you have it, then; Lincoln has been given a moderately major facelift for '57 (although "rump-lift might be a better term in this case!) Is it for better or for worse? It's up to you. •

BODY TYPES FOR '57

CAPRI SERIES	PREMIERE SERIES
Four-door sedan	Four-door sedan
Two-door hardtop	Two-door hardtop
Four-door hardtop (Landau)	Four-door hardtop (Landau)
	Convertible

THE ENGINEERING PICTURE

Exhaust pipe, in an unorthodox vertical position, is pointed out by stick in photo. This may reduce bumper discoloration.

Like many other makes, Lincoln has gone to a paper air cleaner element for 1957. Engine ratings are straight 300 hp.

Driver's Report— THE 1957 LINCOLN

By Ken Fermoyle

IF I HAD BEEN blind-folded (obviously not a very good idea!) I would have had a hard time telling the 1957 model Lincoln I drove from '56 Lincolns I've driven during the past year. Handling, ride, performance—they all felt about the same as on the earlier cars.

This was no big surprise as far as ride and handling are concerned. There have been only minor changes which would affect these characteristics. And there was little reason to tinker in this department either because the 1956 Lincoln was a fine all-around blend of good qualities.

I put the '57 thru just about the same series of handling and roadability checks on Ford's Dearborn test track as I did with the '56 a year ago. My impression was that there is very little difference between the two cars. The changes in shocks and shock mountings to minimize slight wheel movements will undoubtedly show up in a more comfortable ride, as Lincoln engineers claim, but this has little overall effect on the feel of the car.

You can still put the Lincoln thru a fast turn with great ease and less roll than most competitive luxury models can boast ... and I found in '56 that it outshone a lot of smaller cars that might be expected to handle better than one of Lincoln's size and weight. Despite this, there can be no quarrel with riding qualities. Lincoln is undoubtedly one of the top choices for long-distance, high speed touring in its class. You can cover a lot of miles in one at an excellent average speed without driver or passenger fatigue.

I was a little disappointed in the performance of the new Lincoln I drove, however. I didn't expect it to be greatly changed from last year, but I did expect some improvement due to higher compression and other modifications—especially since there was no significant weight increase.

Unfortunately, the only pre-production model available for testing was not in a good state of tune. The Lincoln engineer who worked with me warned me about this before we started. He explained that this particular car had been used extensively for styling pictures (for advertising purposes etc.) and chassis development work recently and the engine boys had not had an opportunity to get the V-8 up to snuff.

For that reason the acceleration figures can't be regarded as representative and we will endeavor to check the performance of a more typical '57 Lincoln as soon as possible. One significant thing, however, was that 0-to-30 mph times were better than last year—4.25 seconds compared to slightly more than 4.5 in 1956. This confirmed what the Lincoln boys had said: that they were shooting for better low and mid-range acceleration and not worrying about the high end so much.

This was further confirmed by the 50-80 mph time of 10.4 seconds (compared with 11.7 for a production '56). From 0-to-60 using low and drive ranges took about 11.5 seconds and 0-to-80, about 19 and a few fractions. I'm not pinning these figures down exactly because we were figuring on an average speedo error based on past experience with Lincolns and there may have been some variation.

I won't be able to say for sure until I've driven and timed production models, but I'll venture a guess that a typical '57 Lincoln in fair tune will turn 0-to-60 mph in 10 to 11 seconds, depending on driver and conditions, with 50-to-80 mph times

WE'VE SEEN what kind of a beauty treatment has been given the sheet metal "skin" of the new Lincoln. As the old adage says, however, beauty goes deeper than that. What's happened underneath?

Mechanically, just as in styling, the 1957 Lincoln has been facelifted—though perhaps to a lesser degree. There are, however, several interesting changes.

In the engine compartment—more power. Naturally. What is slightly surprising is that there aren't any more cubic inches, however. The extra horses result chiefly from a boost in compression ratio to 9.9-to-1 (vs. 9-to-1 in '56), a change in carburetors and minor detail changes concerned mainly with improving breathing. In addition, the full vacuum distributor advance system has been dropped in favor of one using centrifugal advance, and a larger capacity radiator is now used.

The new carburetor is a Carter (four-barrel, of course) with greater air capacity and larger float bowl. Used in conjunction with it is the new paper element air cleaner.

Moving on back to the transmission uncovers some interesting changes. The 12 5/16-inch aluminum torque converter unit has been replaced by a 12-inch steel unit. Lincoln engineers claim this enables them to use the higher engine output to provide better performance in the usable ranges—i.e. Low and middle speed acceleration. Stronger gears are used in the transmission gears also.

Another transmission change explains one reason for the larger radiator. The new radiator has an integral oil cooler thru which the transmission fluid is piped. This arrangement, incidentally, is one we'll be seeing more of in the future as

KEY SPECIFICATIONS FOR 1957

Horsepower	300 @ 4800 rpm
Torque	415 ft. lbs. @ 3000 rpm
Height	60.2 inches
Length	224.4 inches
Width	80.8 inches
Wheelbase	126 inches
Tread	60 inches

engine power and torque rises and new automatic transmissions appear. (Ford uses a similar system for '57.)

Lincoln's suspension continues pretty much unchanged; coil springs and ball joints at front and semi-elliptics at rear. Shocks and shock mountings have been redesigned, however, to minimize small wheel movements and provide a more comfortable ride.

A limited-slip differential will be available as options on new Lincolns this year—as on 1956 Packards. An electric door locking system is also being offered. Other Lincoln accessories include a fuel warning system consisting of a light on the dash wired to flash when you get down to about three gallons of gas. Vent windows which are power operated can now be ordered. The optional Multi-Luber has been improved (now has an automatic cycling device). All normal power-operated equipment and accessories will, of course, be offered again in '57. In fact, Lincoln advertising will speak of the car as the "most automatic car on the road," or words to that effect. •

Fermoyle takes the new Lincoln around a very hard corner on Dearborn handling course. Car still rates as better-than-average.

running within a few tenths of the 0-to-60 figure. I won't even guess at top speed, although it will be more than ample for practically all purposes.

Here are a few other impressions:

Visibility is excellent. All four fenders can be seen from the driver's seat.

Seating position is excellent and getting into the car is as easy as on any car of similar height.

Switches and gages are placed quite sensibly.

The power steering still lacks "feel" to some extent and too many turns of the wheel are required in tight turns.

Interiors are even plusher than last year. Lincoln is right up with the best in this respect.

Some of the new accessories—limited-slip differential, fuel warning signal, electric door locks, improved Multi-Luber etc.—are valuable additions to the optional equipment list.

What does it all add up to? My feeling is that the new Lincoln retains all the good qualities that enabled it to set new sales records in the past years and has added a few more. It has been changed, overall, more than you might expect after the major revamping for '56. As far as styling is concerned, my personal opinion is that some of the sleek, almost classic, good looks of the car have been sacrificed in the effort to make the '57 look like a different car. That's just one man's opinion, however. And I still think that Lincoln is one of the more desirable cars in its class. •

1957 LINCOLN

Front of '57 Lincoln has even more massive look than previous models. Smaller road lamps have been added below headlights. Directional and parking lamps are built into the bumper bar.

THE NEW LINCOLN is a face-lifted and improved version of the completely re-designed 1956 model.

In addition, a four-door hardtop body has been added to the lineup of Capri and Premier four-door sedans, two door hardtops and convertibles.

Engine displacement remains at 368 cu. in. but horsepower has been increased from 285 to 300. A 10.0 to 1 compression ratio, improved manifolding and a new distributor are responsible for the gain.

The car's front end retains the low, wide, predominantly horizontal character of previous Lincolns, but has been altered by the addition of driving or "road" lights beneath the headlamps as a standard equipment item. (Lincoln calls this Quadri-Lite styling.)

The side appearance of the Lincoln has been considerably changed by elimination of the stainless steel spear molding along the bottom of the body panels.

One of the most important of optional accessories for '57 is limited-slip differential gearing which prevents a "free" rear wheel from spinning under conditions of unequal traction. ●

Four-door hardtop in Lincoln's Premier series is a new body type for '57. Conventional four-door sedans have new, thinner door pillars. Although overall length of the Lincoln has been increased, wheelbase remains 126"

Side view with hood erected gives an impression of car's length, lowness, pleasing appearance. "Continental" type spare wheel mounting is preserved.

Rebirth of a classic...

First new Lincoln Continental Convertible in nine years

RESULTING from the joint efforts of the Lincoln Division of the Ford Motor Company and the Derham Custom Body Company, the first Lincoln Continental convertible in nine years recently made its U.S. debut.

It is the Continental Mark II Cabriolet convertible, and continues the classic styling theme of the Continental Mark II hardtop coupe, which is at present being produced in limited numbers by the Ford Company.

The car shown was finished in pearlescent white lacquer, with interior trim of contrasting red and pearlescent white leather, and stood only 57 inches high.

The Lincoln Continental is and has always been a costly high grade, prestige car for discerning buyers.

Pre-war Lincoln Continental convertibles today are still eagerly sought by enthusiasts in the U.S.A., and are still fetching extremely high prices. The long, low styling with accentuated rear deck has always been a Continental feature, and the largely hand built cars have always preserved a long, lean look which has been often copied but seldom successfully imitated. The outside-mounted spare wheel, a distinctive Lincoln Continental feature over the years, has given instigation to a variety of different "Continental" spare wheel kits made by different accessory manufacturers for fitting to various stock automobiles.

Production was strictly limited, and the series eventually faded into limbo in March, 1948. Diehard Lincoln adherents all over the world viewed with sadness what appeared to be the final demise of what many consider the finest car ever made in the U.S.A.

The new Lincoln convertible will meet with much rejoicing, particularly since the traditional classic lines and high grade engineering of the older models have been carefully preserved.

Unveiled at the recent Texas State Fair... the first true Continental convertible built since production of the Lincoln Continental series was suspended in March, 1948. Car is quality built in traditional Lincoln manner, has distinctive styling.

Low slung, dignified, commanding... the Lincoln Continental Mark 11 Cabriolet convertible.

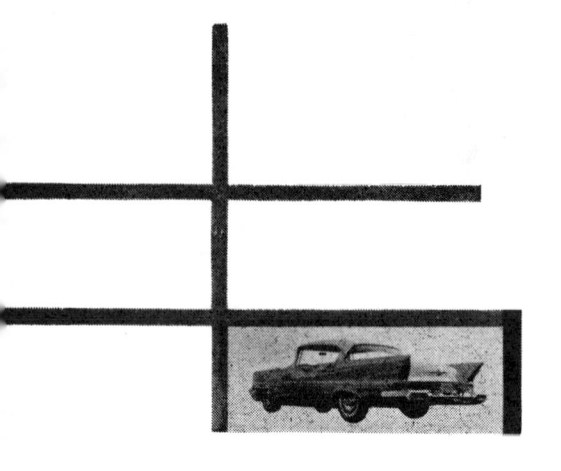

DRAMATIC FIN STYLING, LOWER THAN EVER, AND IMPROVED ROADABILITY FOR Lincoln

What's New?

Major facelift, front and rear, emphasizing dual headlights in vertical pairs, canted fins . . . A four-door "landau" hardtop, new to the Lincoln line . . . More power, smoother ride . . . Optional "power-directed" differential and electric door locks . . . Standard Turbo-Drive transmission, power steering and brakes.

Your Choice

Lincoln's two models, the Capri and Premiere, are the same mechanically, differ on the inside in trim details and quality of fabric. The more expensive (by about $350) Premiere is factory-born with power windows and front seat. If you want a convertible, it must be a Premiere also. All other models are available in either form.

For some reason, the company went to great lengths in redesigning their four-door sedan *in addition to* bringing out the new landau hardtop. Marketwise, Cadillac looked around this year and dropped the "B" post (between the two doors) entirely. Lincoln, on the other hand, fixed up their sedan so that when the windows are closed, you can't tell it from a hardtop.

The overdue landau, a true four-door hardtop, costs more and is mounted on a heavier, stronger frame to make up for lack of support from the body structure. This is also true of the styles that make up the rest of the line, the two-door hardtops and convertible.

Fine car choice (perhaps we should say high-priced choice as practically every make qualifies as "fine" nowadays) has narrowed down to three, for the forthcoming new Packard won't be in this league. Whether you choose a Lincoln, Imperial, or Cadillac seems to be either a personal or business matter between you and yourself. All three have reached present day ultimates in both luxury and performance.

Lincoln Power

This year's engine seems to have needed very little attention to remain competitive. We'll qualify that last word by stating that when power gets up to 300 horse, the car is capable of winning races that shouldn't be run on public highways. An extra 25 or 50 horsepower is a purely academic bonus, nice to have but not worth paying for if usage planned is normal. We don't think that Lincoln will lose any sales because they don't *advertise* tops in horsepower.

Engine improvements revolve around a 10 to 1 compression ratio for better economy on premium fuels and a redesigned carburetor that is more resistant to hot-weather loss of performance. The more you insulate an engine compartment, the worse the "heat-soak" problem becomes. Gasoline can literally boil in the carburetor bowl, and Lincoln is one of the first to recognize this. A supply of dry-ice might work, but an easier way is to provide for a larger volume of fuel in the float chambers so that you can't boil dry and be unable to restart a hot car.

Other commendable carburetor feature is baffles to prevent fuel starvation in one bank of cylinders, over-richness in the other, during high-speed cornering. Like most '57 makes, Lincoln now uses a paper-pack air filter for better filtration and easier maintenance. A new distributor completes the list of engine changes.

The Turbo-Drive (similar to Merc- and Fordomatics) transmission has been worked around internally for better durability, but operation is unchanged. You will note a new radiator, mounted like a kangaroo carrying its young on the main water radiator (see photo). This is to cool transmission oil more efficiently.

Lincoln on the Road

From 1952 through 1954, no one seriously questioned Lincoln's position as the top road car, because for all three of those years in a row there was a 2000-plus mile road race, called the Carrera Panamericana, which Lincoln won.

When the 1956 Lincoln came along, with its new, softer-sprung chassis and elongated lines, enthusiasts were inclined to doubt the breed. The argument is still going on, for the 1957 Lincoln is also of the later mold. Unfortunately, there is no longer a Mexican Road Race to prove its maker's claims that the car will stick with and maybe pass any Lincoln ever built on any kind of a road. In any case, you'll enjoy the boulevard ride.

Inside Your Lincoln

Owners since 1951 will note a familiar instrument panel, trademarked by the big horizontal speedometer. Controls are all conveniently placed, and have benefited from Ford Motor Co.'s exhaustive safety research.

Figures haven't been tallied up this year, but we believe Lincoln is still near the top in glass area, and therefore visibility. It is too bad the company decided to compromise on their so-called quadra-headlight system, which in reality is just a pair of regular sealed beams augmented by foglamps. The installation was obviously designed for the real thing; just as obviously there will be a running change-over later in the year.

Useful interior gimmicks include the electric door latches with a warning light should a door come unlocked, and an outside rear view mirror that can be adjusted from within. Upholstery, needless to say, is luxurious.

Why Buy?

Award-winning styling made even more beautiful . . . Smooth, quiet power from a well-proved engine . . . Meticulous attention paid to insulating car and occupants from road noise and shock . . . For the first time, complete range of body styles.

FOUR HEADLIGHTS can fit with no major change.

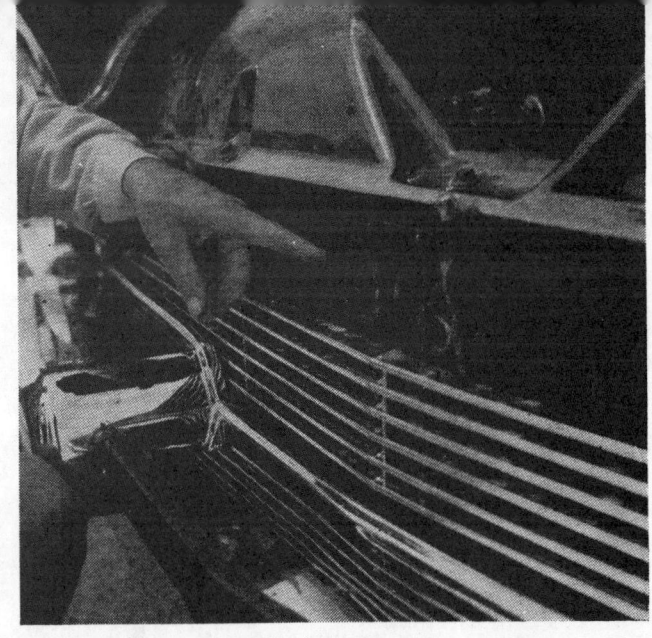

REGULAR SEALED-BEAM HEADLIGHTS

FOGLAMPS—TO BE REPLACED BY DUAL HEADLIGHTS LATER ON IN MODEL YEAR

OIL COOLER insures smooth sailing for torque converter.

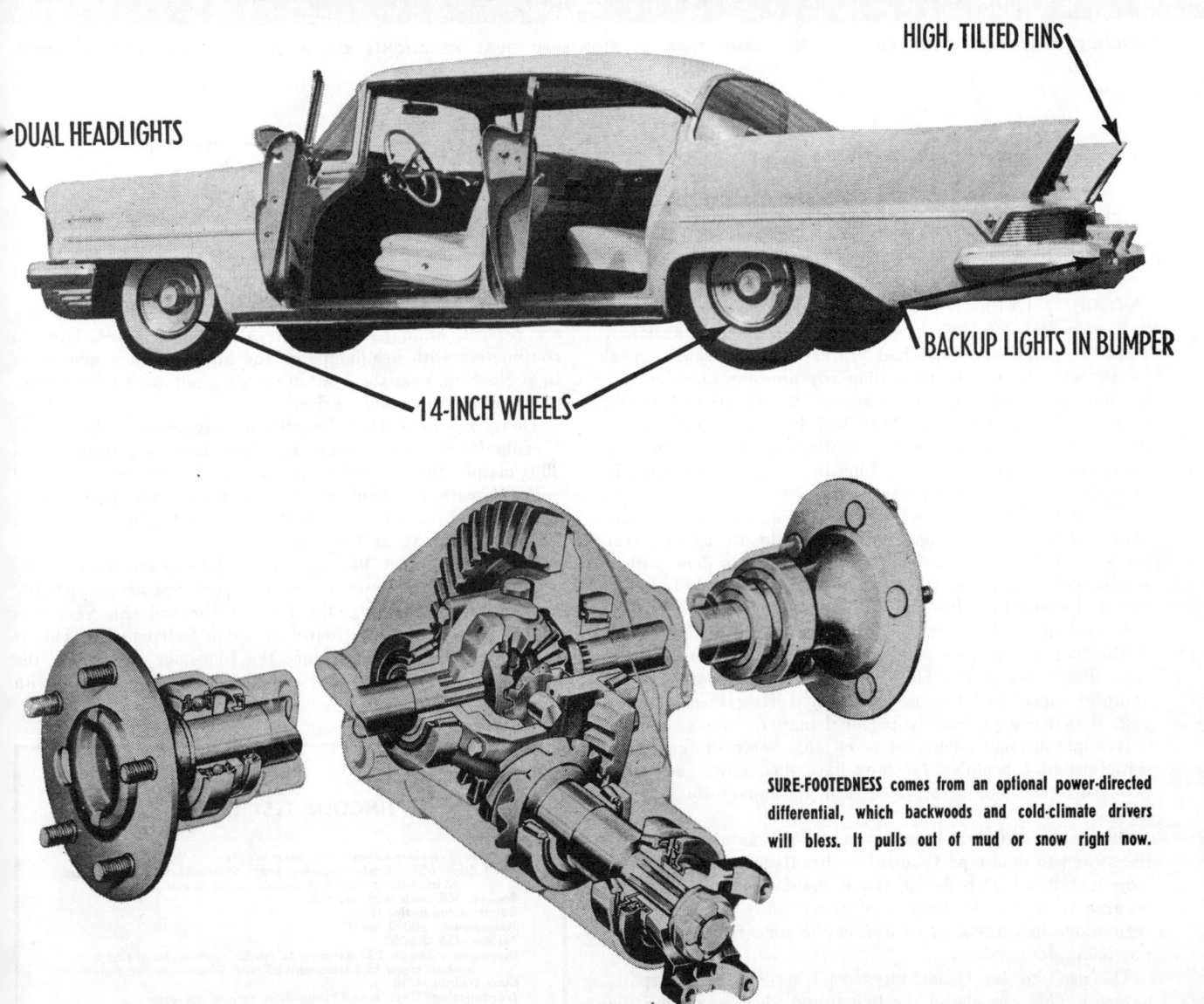

DUAL HEADLIGHTS

HIGH, TILTED FINS

BACKUP LIGHTS IN BUMPER

14-INCH WHEELS

SURE-FOOTEDNESS comes from an optional power-directed differential, which backwoods and cold-climate drivers will bless. It pulls out of mud or snow right now.

SPECIAL 1957 SHOW ISSUE!

LINCOLN HENCEFORTH WILL BE ABLE TO CLAIM THAT IT PIONEERED DUAL HEADLIGHTS ON A STANDARD PRODUCTION MODEL.

LINCOLN ROAD TEST

NOBODY EXPECTED Lincoln to make major changes in its models for 1957. After all, the 1956 Lincolns were new from the ground up—and had proven they had sales appeal by attracting more customers than any previous Lincolns.

Thus, when the new models appeared, they created a mild surprise. Not that they actually had been changed a great deal; they just *looked* quite a bit different. Without making many real changes in the car, Lincoln designers succeeded in changing its character significantly.

The most obvious revision was in rear quarter panels. Lincoln had been one of the last of the holdouts against rear fender fins. When the switch was made, it was done with a vengeance! And it was this change to high, canted-blade fins which changed Lincoln's character most.

In adding the fins, Lincoln lost some of the distinguished, distinctive, almost neo-classic look it had in 1956. At the same time it gained, too. The 1957 Lincoln has a sportier, more youthful appeal. What it lost in dignity it has gained in added zest. Which you prefer is a personal matter.

It might be noted here, however, that some of the initial criticisms of Lincoln's 1957 look have died down, now that people have become accustomed to it. So apparently it grows on you.

Other than addition of the fins, major styling alteration was the switch to dual-type Quadri-Lite headlamps. And this was more a styling than truly functional change—mainly, perhaps, because there was so much confusion legally about lighting regulations in various states during the time the '57 Lincoln was being designed.

This isn't to say Quadri-Lites don't serve a real purpose, however. This was one of the first things checked during the road test of the Premiere Landau four-door hardtop used for the purpose. It was found that the auxiliary driving lights—smaller of the two sets of lamps mounted horizontally above the grille—are very helpful under certain conditions.

At dusk, for example, when it's not dark enough so that you need headlights but do require some illumination, they are perfect, although this is illegal in some states. Used in conjunction with headlamps on the highway, they give extra light close-up, near the front of the car, and don't add to glare problems of on-coming drivers.

There are even fewer significant engineering changes in Lincoln for '57. Horsepower has been increased from 285 to 300, mainly thru a higher compression ratio and improved engine breathing. Displacement remains at 368 cubic inches. Compression ratio increase to 10-to-1 helped raise torque from 402 to 415 lbs./ft. at 3000 rpm.

The power boost has apparently been cancelled out by weight increases, however, because performance is virtually unchanged. If anything, the Landau checked this year was slightly slower than the four-door sedan tested in '56. This is understandable, however, since the four-door hardtop is the heavier of the two body styles and it was literally loaded with

---

LINCOLN TEST DATA

Test Car: Premiere Landau four-door hardtop
Basic Price: $5221 (includes power seats, windows, steering, brakes and automatic transmission; all standard equipment)
Engine: 368-cubic-inch ohv V-8
Compression Ratio: 10-to-1
Horsepower: 300 @ 4800
Torque: 415 @ 3000
Dimensions: Length 224.6 inches, width 80.3 inches, height 60.2 inches, tread 58.5 front and 60 rear, wheelbase 126 inches
Curb Weight: 4710
Transmission: Three-speed Turbo-drive torque converter
Acceleration: 0-30 mph 4.2 seconds, 0-45 mph 7 seconds, 0-60 mph 11.7 seconds
Gas Mileage: 11 mpg average
Speedometer Corrections: Indicated 30, 45 and 60 are actual 30, 47 and 60.5 respectively.

140

CRISP MODERN LINES have been sharpened as Lincoln finally yielded to tail fins. The test car was a Landau four-door hardtop, a new body type for the make in '57. Trunk area is huge.

power equipment and accessories. In fact, it weighed a little over 5000 lbs., including driver and passenger, during acceleration runs.

One thing acceleration runs did prove and that's that this big, heavy car is pretty nimble in the normal highway passing speed ranges. Several runs from 50 to 80 mph netted an average of 11.2 seconds. This is not blinding, of course, but it's par for the course and shows that the Premiere would move well after it had a chance to pick up momentum.

Although Lincoln doesn't have the clear margin of superiority over several rivals in its price class this year as in '56 so far as handling is concerned, it's still right up with the best in this department. For a car of its bulk, it's remarkably agile. Don't misunderstand; you won't think you're tooling a sports car when you're behind the wheel of a Lincoln.

What you *will* get is a remarkably stable ride under most normal—and some abnormal—conditions for a car of this size and type. You don't get the tremendously comfortable ride and impressive length a Lincoln offers along with maneuverability and cornering qualities of a small car. Nor should you expect to. For really solid comfort under highway cruising conditions, however, along with stability and handling qualities more than adequate for the demands of most luxury-class car buyers, Lincoln is hard to beat.

Going back to acceleration and performance, it should be pointed out that the car tested was barely broken in when these checks were made. It had only 550 miles on the odometer when picked up and just under 700 when acceleration tests were made. It's reasonable to assume that times listed here would be improved after the car was loosened up a bit more.

One of the accessories most appreciated on this car was the outside rearview mirror which could be adjusted from inside the car. It could be controlled by the driver just like a side-mounted spotlight. This is not only convenient, but is a definite safety feature since it permits mirror position to be changed quickly and easily as needed to provide proper vision.

Another safety feature which will have special appeal for Lincoln owners with small children is the electric door locking system. Pressing a button automatically locks all four doors and keeps them locked until they are unlocked manually or the driver shuts off the engine and opens his door.

Not so safe were the brakes. After a period of admittedly hard usage, they began to fade rather badly. They returned to normal after a short cooling-off period, but something like this could be embarrassing under certain conditions—on long mountain down-grades, for example.

From a quality standpoint, the test Lincoln was about average for the industry. The finish was good and panels were well fitted, but a critical eye could find a few ripples in the sheet metal—particularly along the inside surface of rear fender fins.

Interior trim was excellent. Upholstery was a combination of leather and metallic-thread brocade. It looked not only handsome, but practical from a wear and cleaning standpoint.

Visibility is good in all respects in the new Lincolns. The driver can see all four fenders from his seat—and darn near anything else at front, side or back. Instruments are legible and easy to reach, although the ash tray is mounted to right of center of the dash so the driver has a long reach to get to it.

On the whole, while Lincoln hasn't changed as much for '57 as some of its competitors, it still has an awful lot to offer buyers who are looking for a car in its price class. •

REAR DOOR hinges at front edge which affords easy entrance and exit and also helps prevent door blowing open while moving. Section of top opens with door, as shown in photo at right.

ADJUSTABLE MIRROR can be manipulated from inside, something new for Lincoln although it has been used previously. A really tricky device enables driver to lock all doors electrically from dash.

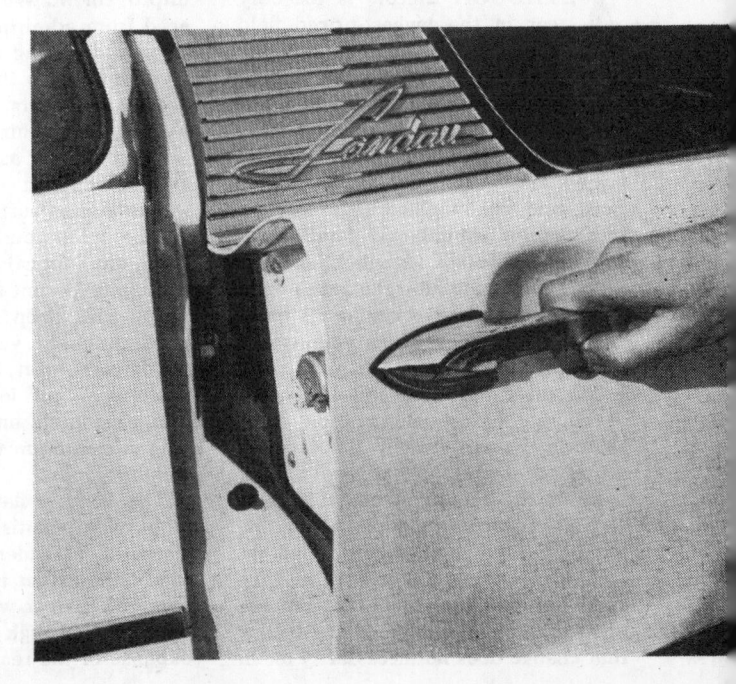

CAR LIFE *1957 CONSUMER ANALYSIS*

LINCOLN

By JAMES WHIPPLE

ALTHOUGH Lincoln is the only car in the highest-priced field that has not been completely redesigned for '57, it rates second to none as a top quality, luxury automobile.

The Lincoln was completely redesigned last year — suspension, frame, body and engine. The 1957 version incorporates a number of detailed improvements plus enough "face-lift" styling changes to alter the appearance of the car without changing its basic character which remains unmistakeably Lincoln.

The most noticeable styling change, of course, is the substitution of a smoothly swept fin for the tapered contour of last year's rear fender. Also, there is the raising of the side trim and the incorporation of twin headlamps into the grille and front end.

The overall effect is smooth, elegant and pleasing, but we feel that Lincoln this year is an example of the theory that change does not necessarily mean improvement. We still haven't recovered from admiring the tremendously improved styling of the 1956 Lincoln as compared to the '55 model.

One of the first, happy impressions we received behind the wheel of the '57 Lincoln was, as last year, the really excellent vision which is especially welcome, and surprising, in a big car.

The wraparound windshield, free from any optical distortion, is wide enough to permit an excellent field of vision and deep enough to provide good overhead vision for the tallest of drivers even though the seating position is not low-slung. Unusually thin windshield and door posts block only a minimum of the driver's side vision.

The hood is flat and slopes gently to provide surprisingly good forward vision. The fender fins act as guides to aid the driver in locating the "corners" of the car when he has to look backward through the wide, unbroken expanse of the rear window.

Parking the '57 Lincoln will be easy as long as you have the space. The car has been lengthened this year and is now a whopping 224 inches from one massive bumper to the other.

As we examined the interior of the Lincoln, we came to the conclusion the car was unbeatable from a quality standpoint. Only those who measure the Lincoln against the Rolls, Continental or Mercedes are liable to find it second best as far as finish and workmanship are concerned.

Workmanship on the exterior is also really superb. Everything on the car fitted perfectly and the finish of the paint is a wonderful thing to behold after looking at the acres of rumpled metal and "orange peel" paint jobs found on so many of even the medium-priced cars today. Lincoln has improved in finish and quality by about 500% in the past two years.

The upholstery was soft on the surface and desirably firm underneath, like a fine innerspring mattress. In the

Price range (Factory list price)
$4,576 (Capri two-door hardtop)
to $5,300 (Premiere convertible)

LINCOLN
is the car for you

if... You appreciate outstanding quality of workmanship in a mass production luxury car.

if... You like the utmost in a quiet well-behaved and well-controlled motor car that's easy to operate.

if... You are one of the people who like their car interiors roomy and comfortable with a maximum of vision and ease of entrance and exit.

if... You want modern good looks combined with an air of quiet conservatism in your car's styling.

LINCOLN SPECIFICATIONS

ENGINE	V-8
Bore and stroke	4 in. x 3.66 in.
Displacement	368 cu. in.
Compression ratio	10:1
Max. brake horsepower	300 @ 4800 rpm
Max. torque	415 @ 3000
DIMENSIONS	
Wheelbase	126 in.
Overall length	224.6 in.
Overall width	80.3 in.
Overall height	60.2 in.
TRANSMISSION	Turbodrive

CAR LIFE 1957 CONSUMER ANALYSIS

Headlights in vertical pairs, new grille styling and wraparound bumpers dress up the Lincoln for 1957.

A new addition to the Lincoln line, Landau four-door hardtop is available in both the Premiere and the Capri series.

Hardtop coupe is also available in both series. Engine under Lincoln hood has higher horsepower, compression ratio in '57.

car we drove, a convertible, the combination of leather and tightly-woven cloth, was flawless. All the trim fit perfectly, the instruments and controls functioned smoothly, and the paint and chrome was top quality.

The instrument panel seemed a bit gadgety with its multiple groups of accessory switches and a big spread of heater indicator dials and controls that tend to attract, or should we say distract, attention from the speedometer which is small and not too legible.

The reason for the latter problem is that the markings of the 10 miles per hour speed increments are white lines on the dial glass, while the needle is a thin, white pointer behind the glass which offers no helpful contrast and gets "lost" among the white lines that fan out to the row of numbers on the top of the dial.

When fully equipped, the Lincoln becomes a paradise for the push-button fans with hidden electric motors buzzing behind every panel. At the driver's left hand for example are four electric window switches to raise and lower the windows, two more to open and close the ventilator panes, plus an additional pair for regulating the six-way power seat.

The radio "tunes itself" with a tiny motor, while another one drives the antenna up and down. The convertible's top is operated by electro-hydraulic pump, while two more motors in the windshield header bar operate to screw down the top bow when it falls in place.

Heater and defroster controls have separate blowers and temperature regulation and also permit fresh air or recirculated air for heating. The heater is located under the seat and is so noiseless that we hardly believed it was operating until we realized that eight degree outdoor temperature air was warmed to a comfortable 65 degrees, and this in a convertible!

If silence is a measure of luxury, and we reluctantly admit that it seems so in U. S. cars, Lincoln is just about tops. Full-throttle operation is the quietest of any car, save Lincoln's smooth cousin, the Continental.

Not only is Lincoln's Turbodrive second to none for quiet performance it's tops in smoothness of operation. There's no balky operation during warmup even in extremely cold weather.

We were never able to determine the shift point between intermediate and high ranges on the test car, and only under full throttle did a slight surge betray the shift from low to intermediate.

For the average driver, it would be impossible to tell the difference between Lincoln's Turbodrive which combines three automatically-shifted forward speeds with torque converter and a pure torque converter transmission, except for Turbodrive's superior acceleration and better gasoline mileage.

Lincoln's power steering setup, (it's standard on all models), is not a necessary evil as in some cars. As we drove the Lincoln, we were never conscious of power steering "action." There was no "off-and-on" sensation of the unit "taking over" after a certain amount of pressure was exerted on the wheel rim.

Nor, on the other hand, was the steering assist so "helpful" and continuous that the slightest pressure on the wheel sent the car into a swerve. Two very welcome features of Lincoln steering are its relatively fast action (3.7 turns from lock to lock), and its built-in caster action which tends to return the wheel to the straight-ahead position after turning.

Lincoln's ride, excellent in '56, is even better now. All the harsh feel is blotted up from the roughest surfaces,

Lincoln exhaust outlets are hidden under wraparound bumper. Fins containing pyramid tail-lights are canted outward.

and irregular roads with dips, bumps and high crowns are taken in stride with a gentle, pillowy fore-and-aft oscillation of the body. This never becomes uncomfortable because shock absorber control is just about perfect.

There's a fair amount of heel-over in taking sharp turns at better-than-average speeds, but there's never any uncontrolled wallowing of plowing no matter how fast you barrel into a turn. Although Lincoln's suspension was completely redesigned last year there's still more dip in severe braking than on some of the new cars where it was specifically "designed out." The over all effect of the Lincoln on the road is that of an extremely softly sprung car yet one which is miraculously stable and well-controlled.

Performance of the '57 Lincoln was about the same as last year, brisk and powerful but not world beating. Acceleration from rest to 60 mph took 11.5 seconds, while 60 to 80 mph in 10 seconds proved the car a good high-speed performer.

SUMMING UP: Lincoln is a very well-balanced luxury transportation package that offers incomparable comfort, top quality workmanship throughout, plus great ease of control, with smoothness and quietness second to none. •

LINCOLN CHECK LIST
5 CHECKS MEANS TOP RATING IN ITS PRICE CLASS

PERFORMANCE	Very powerful, but not quite the top performer in its class. Smooth and quiet under all conditions and at all speeds. Turbodrive transmission is the smoothest and most efficient in the industry.	✓✓✓✓
STYLING	Lincoln has smooth, well-balanced lines which minimize any possibility of a "bulky look." New, finned rear fenders have distinctive yet conservative lines.	✓✓✓✓
RIDING COMFORT	Lincoln rates tops of all cars for absorption of surface vibration and small bumps. Car has gently oscillating gait on extremely rough going at high speeds. Control of sway is excellent, although there is considerable lean in cornering.	✓✓✓✓
INTERIOR DESIGN	Headroom, legroom, visibility and comfort of seating positions both front and rear make Lincoln the outstanding "living room on wheels" of the luxury car field. Instrument legibility is a bit below average and controls are cluttered and awkward to reach behind steering wheel.	✓✓✓✓
ROADABILITY	Although not as roadable as the Lincolns of Mexican Road Race fame, the current Lincoln chassis offers excellent roadability coupled with an extremely soft ride. Car remains very stable under all but extreme conditions.	✓✓✓✓
EASE OF CONTROL	Lincoln's Turbodrive transmission with its three speeds and butter-smooth shifting gives the car instant response in traffic, smooth holding action on downgrades. Power steering is an excellent compromise between partial and full assist types.	✓✓✓✓✓
ECONOMY	Lincoln buyers, like those of other high-priced cars cannot expect budget-car economy, but Lincoln will do better than some of its competition.	✓✓✓✓
SERVICEABILITY	Lincoln has the problem of a big, intricate engine covered with a lot of accessory units. Power lubrication system at extra cost will eliminate some service stops.	✓✓✓✓
WORKMANSHIP	Here is where the Lincoln takes the lead. As the Ford Motor Company's strong bid to capture a bigger slice of the luxury car market, the Lincoln is as flawless as any U.S. automobile with the exception of the Continental.	✓✓✓✓✓
VALUE PER DOLLAR	Although it doesn't have quite the prestige or very favorable low depreciation of another luxury car, Lincoln offers 100 cents worth of top quality automobile for the dollar.	✓✓✓✓

LINCOLN OVERALL RATING... 4.2 CHECKS

1958 LINCOLN

WILD REAR END, yet probably the most attractive view of the new styling, is exaggerated further by odd camera angle. The top photo is the standard Lincoln, the bottom right is the Mark III. Influence here was La Tosca dream car, covered exclusively in the December 1956 issue of MOTOR LIFE.

INSTRUMENT CLUSTER has Continental heredity, rather than Lincoln. Steering hub has been recessed further than before.

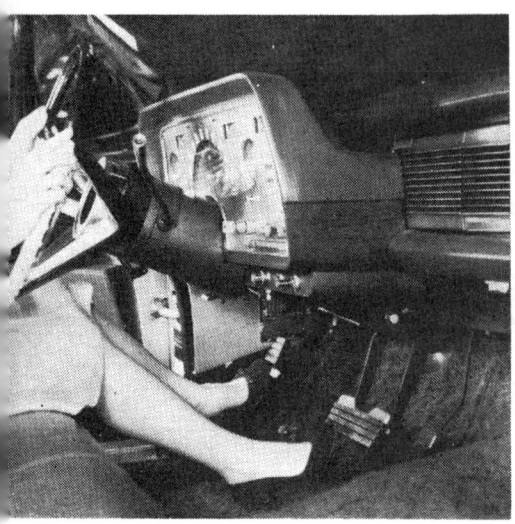

Most revolutionary of all Detroit's 1958 models in both styling and engineering—including a new version of the Continental called the Mark III

LINCOLNS and Continentals are not only all-new for 1958, they are the most drastically changed cars of the year. They are completely new from the ground up—literally, since even the tires have been changed.

They offer entirely different styling from '57, sweeping engineering changes and aren't even built in the same way.

(It's been an open secret for months that Lincoln would switch to unitized construction for 1958, in fact had built a plant outside Detroit designed just for that purpose.)

Lincoln's model lineup has been expanded by addition of four cars in the Continental Mark III series. Mark III models are a two-door hardtop, four-door hardtop, four-door sedan and a convertible. All except the convertible are offered in the Premiere and Capri series, making a total of 10 models available in 1958.

Continentals are basically similar to Premieres and Capris, share the same body, engine and have exactly the same exterior dimensions. They have a different grille, their own distinctive interiors and are the only cars in the Lincoln lineup which offer a reverse-slant, power-operated rear window. Prices, not available at press time, will be "substantially lower than the Mark II."

All 1958 Lincoln and Continental models feature aluminum grilles and compound curved windshields. Dual headlamps are standard equipment and they are mounted in a unique fashion —they are set one above the other at an angle in front fenders.

Straight, clean lines and broad, flat sheet metal panels with extensive use of sculptured effects characterize 1958 Lincoln.

The cars are much bigger than in 1957—and look it. Wheelbase of both Lincolns and Continentals has been increased from 126 to 131 inches. Overall length is 229 inches, making them the longest cars in the industry. Overall height is 56.5 inches, almost four inches lower than 1957 Lincolns and just a half-inch higher than Mark II Continentals. The switch to unitized construction made it possible for Lincoln to cut height with virtually no sacrifice in head room.

All models have a suspension system using steel coil springs as standard equipment. Trailing arms take up the driving torque at rear and a track tied to axle housing and chassis locates the rear axle laterally.

This design makes it easy to install the air suspension system which is optional on all models. Air springs are simply substituted for the coil springs and the compressor, tanks, leveling and check valves and air lines are installed. Only real change required to fit the air suspension system is relocation of front shocks, normally mounted in the center of the coil springs, to a point just back of front air springs.

A new 375-hp, 430-cubic-inch V-8 is standard in all three series. (This engine is almost identical to the larger 1958 Mercury engine.) Despite its greater size and horsepower, the engine is 17 lbs. lighter than the 1957 Lincoln V-8 and is designed to permit faster, easier servicing. Compression ratio is 10.5-to-1 and a four-barrel carburetor is standard. (Lincoln apparently does not plan to offer a triple two-barrel carburetor setup like Mercury's, however.)

Lincolns and Continentals are the only U. S. cars in their classes which use unitized construction. Bodies and frames are not separate components as in conventionally-built cars. Body and underbody components are welded together in one integral unit. This not only eliminates squeaks and rattles caused by loose body bolts, but increases torsional rigidity and thus makes for better door and panel fits.

An important quality feature is a body dip process which protects against rust. Entire bodies of Lincolns and Mark III's are immersed in a tank containing rust-resistant primer to a height of 28 inches above the rocker panels.

This coats the cars completely in the area most vulnerable to rust and corrosion—and does it thoroughly, since primer gets into areas which can't be reached using the normal spray.

Power brakes are standard equipment for all models and, because 14-inch tires and wheels are now used, brake drum diameter has been reduced from 12 to 11 inches. Front lining width has been widened one inch and rear width widened one and one-half inches, however, so there has been an increase of 79 square inches in effective brake lining area.

A new cowl-type air conditioning and heating system is offered as an option for all models. It is the only front-located air conditioner in the industry offering direct cooling of the rear seat area. Ducts route cool air from the conditioner through front door panel ducts to the rear seats.

Lincoln has been encouraged, but not satisfied, with sales in 1956 and 1957. The new body style introduced for '56 brought a resurgence for Lincoln in the luxury car class and company planners have made a large bet that the new models just introduced will increase volume even more. It will be a serious blow if this doesn't happen—but Lincolns and Continentals seem to be just the type of big, impressive automotive package luxury car buyers appear to want. •

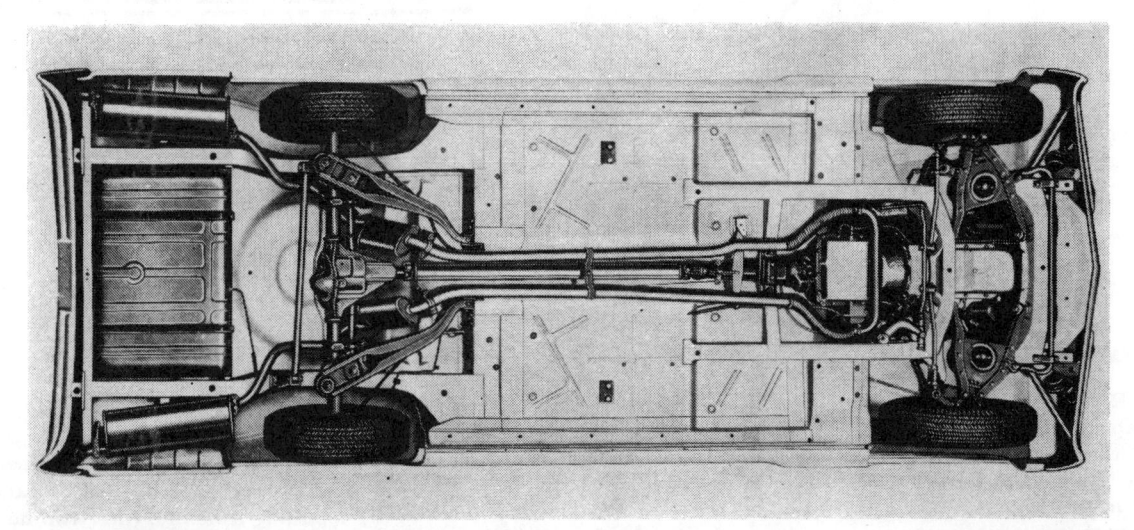

DRASTIC CHANGE, and most fundamental one, is the use of unit body, not entirely new for Detroit but comparatively rare. Move to a new plant permitted the step, which will be shared by the 1958 Thunderbird.

MARK III is a Lincoln Continental, since separate Continental production ceased last May. Front grille is chief styling difference from the standard Lincoln. Other than that, it resembles the Mark II.

STANDARD LINCOLN, along with the Mark III, is the only '58 model to retain a vertical dual headlight treatment. Almost every panel on the body has deeply sculptured styling treatment.

CAR LIFE *1958 CONSUMER ANALYSIS*

LINCOLN

By JIM WHIPPLE

FOR THOSE OLDTIMERS who remember Lincoln, along with Packard and Pierce Arrow, as one of the finest automobiles ever produced in this country, 1958 will be a year to remember.

Lincoln is now back at the top of the heap. For the first time since World War II, Lincoln shows a clear superiority to competition in almost every area of comparison. This doesn't imply that the competition isn't very very good indeed. In one or two features the '58 Lincoln is equalled, perhaps even excelled.

But in the overall package that adds up to owner satisfaction, driver-ease and passenger comfort, the Lincoln is a really first class automobile.

The comeback trail has been a long one for Lincoln, but regaining the heights of automotive excellence is even more of a triumph because of the nearly two decades during which Lincolns were something less than what they had been or what they might have been. 1939 was the last year of production of the magnificent old Lincoln, a twelve cylinder giant with a locomotive-like chassis on which some of the best custom and production coachwork in the world rested with complete unconcern for any limitations of height or weight.

After that, Lincoln became merely a De Luxe version of the medium-priced Zephyr, a basically well-engineered unit body and chassis with strictly "Mickey Mouse" components of engine and running gear attached.

By 1956, Lincoln was at last in the groove and on the way to the top of the prestige market with a long, low car which had a very high level of style and quality control. The '56 Lincoln looked long and massive, as did the Lincolns of the '30s, but it was also wide, low and sleekly styled. Riding was considerably softer, somewhat to the detriment of handling, and power output soared to a very healthy figure.

So did Lincoln sales, much to the joy of management, which then sat down and planned the present car.

To say that I like the '58 Lincoln is putting it mildly. After two hours behind the wheel on the Ford Motor Company's Dearborn proving ground, the only disappointment I felt stemmed from the fact that I was financially unable to place an order. It's one of the most totally satisfactory automobiles I've ever driven or ridden in.

What makes it that good? If I had to answer that question in one word I'd say simply, "Engineering!"

For once the styling tail has not wagged the engineering dog to the disadvantage of the car owner. The stylists and engineers who planned the '58 Lincoln had to have a highly developed sense of teamwork to have designed a car as practical and as well styled.

The car's integral frame and body construction spells out many advan-

Lincoln has emerged in 1958 as an all-new car. Distinctive exterior styling takes a cue from the late Mark II, and emphasizes long, low look with lavish use of "sculptured" steel.

LINCOLN is the car for you

if... You feel that a top prestige car should be superior in comfort, performance and quality.

if... You want the very real advantages of integral body-frame construction and a car of unexcelled passenger comfort.

if... You are looking for the ultimate in riding comfort, interior room and quality of workmanship, in a car that's easy to handle.

LINCOLN SPECIFICATIONS

ENGINE	V-8
Bore and stroke	4.30 in. x 3.70 in.
Displacement	430 cu. in.
Compression ratio	10.50:1
DIMENSIONS	
Wheelbase	130 in.
Overall length	229 in.
Overall width	81 in.
Overall height	56.5 in.
TRANSMISSIONS	Turbodrive

CAR LIFE 1958 CONSUMER ANALYSIS

Continental Mark III is distinguished from rest of line by cross-hatched grille and chromeless side panels. Rear window disappears à la Turnpike Cruiser.

tages. It enables a lowering of the overall silhouette by four inches (from 60.5 to 54.5) with no loss of headroom or seat height. It also permits a low, smooth floor with a minimum of obstacles. With the body itself serving a structural function, the passenger compartment can be designed inside the frame instead of on top of it.

The Lincoln body was clearly designed with a great deal of thought to passenger comfort and driver convenience. It's tremendously wide, and both front and rear doors open wide and allow for easy exit and entrance. Although there are other cars longer than Lincoln's 329 inches, there are none with greater back and front seat leg room. Even with the front seat moved all the way back there was more leg room than in any other six-passenger car.

Vision is also terrific. The windshield is wide, deep and smoothly wrapped around, with only slight distortion in the corners. Lincoln engineers Alex McClaren and Dan Wirtz informed me that windshield area alone was over 1700 inches. The hood is wide but much flatter than previous Lincolns and has a definite downslope. As a result, view of the road is excellent.

Side windows are high, as the roof has no wrap-over and windshield corner posts are narrow, offering the absolute minimum of blind spots in any U.S. car to date.

The trunk space is very large and surprisingly high for a car set so low —another result of integrated body and frame construction.

Lincoln's instrument panel is well arranged and has commendably little unnecessary ornamentation, once again a happy compromise between engineering and stylists. Heating and ventilating system is operated by one convenient illuminated dial control which governs an electric motor that does the work of opening and closing dampers and valves.

Control of Lincoln's three-speed and torque converter transmission is by a set of illuminated pushbuttons located in the lower left end of the instrument cluster area. It took me just a few "passes" to become accustomed to the new location, which is far more convenient than any steering column control handle.

On the road the new Lincoln is almost pure pleasure. The big difference doesn't become evident while you are rolling on absolutely smooth pavement. Then, you notice only the excellent vision and extremely quiet operation of the 430-cubic inch engine.

The Lincoln is so smooth, quiet and spacious that it doesn't give the impression of a "hot" car. My stopwatch told a different story, however. From 0 to 60 mph the car hustled in 9.5 seconds, compared with the 11.5 of the '57 Lincoln. The weight of this new car with four men aboard was well over 5500 lbs. That's no mean hauling, even for 430 cubic inches.

At high speeds the advantages of Lincoln's new integral body and frame really became evident. At 100 mph on concrete pavement there is no quiver or shudder as the tar strips belt the tires. With the windows closed we were able to carry on a conversation in normal quiet voices. At this speed, which is high cruising for the Lincoln, top speed is a good 20-25 mph up the dial, the car was rock steady, with no rolling or swaying.

It was on the rough stuff; gravel, cobblestones, brick, and potholes that the '58 Lincoln really showed us what a balanced engineering job could do. The new suspension system is all coil with the rear axle positioned by trailing links against fore and aft movement or braking and driving torque, and a track bar maintaining precise lateral position.

As a result of that, plus the rigidity of the integral body-frame, the wheels are precisely positioned at all times, and there is little tendency for the rear end to perform a hula dance. I made a dramatic comparison with the

Canted dual headlights are unique in the industry. Line is followed on bumper and front fender panels.

Rear deck of '58 Lincoln culminates in bold treatment of bumpers and light clusters, making an integrated design.

'57 Lincoln on a stretch of dirt road pocked with 4-inch-deep by 10-inch-wide washboard.

At 40 miles per hour, the '57 Lincoln, a four door sedan, shook so badly that door frames moved visibly against the center pillar. The entire car slewed and yawed right and left so badly that I could barely keep it on a straight course, and the fenders flapped against the hood side like castanets.

A few minutes later I drove over the same road at the same speed in the rain in the '58 Lincoln, a Premier two-door hardtop. There were virtually no rattles around the doors and windows, although the wheels pounded the road at terrific force. The car was under absolute control at all times, even though the wheels were bouncing from highspot to highspot.

Throughout there was no bobbing or swaying. Only the previous drive in the older Lincoln, itself an exceptionally comfortable car, proved to me how bad the road.

On all types of road the story was the same, exceptional control and absence from jiggling vibration or pitching.

Summing Up: The '58 Lincoln is an exceptional car, offering more room, comfort, performance, quality and driving ease than any other automobile currently produced in America. ●

LINCOLN CHECK LIST
5 CHECKS MEAN TOP RATING IN ITS PRICE CLASS

Category	Description	Rating
PERFORMANCE	Lincoln's performance is very powerful, particularly for a car of its weight. It equals anything in its class and outperforms many light cars.	5
STYLING	Lincoln's styling is impressive, elegant and modern with a happy absence of unnecessary trim. Most important, styling never interferes with comfort, convenience or driving ease.	4
RIDING COMFORT	In riding comfort, the Lincoln surpasses one of its rivals by a considerable margin and just shades the other. The total comfort picture—i.e., silence, freedom from rattles and vibration as well as ride comfort—leaves Lincoln with no equal.	5
INTERIOR DESIGN	In this category, the '58 Lincoln has no real competition, for in ease of entry and exit, seating position, headroom, leg room, all-around vision and convenience of control the car is tops.	4
ROADABILITY	Lincoln surpasses one of its rivals in roadholding and handling, and equals the other. The car has desirable understeer, considerable tire squeal, some lean, but steady and predictable behavior in high speed cornering.	4
EASE OF CONTROL	New location of transmission control and a new and very good power steering system which gives good "feel of the road" makes Lincoln equal to the best.	4
ECONOMY	Pennywatchers aren't going to be shopping for a car in Lincoln's $5009 price range, so the expected 15 miles per gallon should be no detriment to purchase.	4
SERVICEABILITY	An improved engine design should make necessary service operations easier. The integrated body and frame will eliminate much body bolt tightening as well.	4
WORKMANSHIP	Here again, the '58 Lincoln seems to have a distinct edge — as did the '57 — over the other cars in its class. Interior upholstery and coachwork are particularly good.	5
VALUE PER DOLLAR	For the luxury-car buyer, the '58 Lincoln should be just as good as a basic buy as the most popular car in the price field. Quality is high enough to insure public acceptance, and lower depreciation should follow.	5

LINCOLN OVERALL RATING...4.4 CHECKS

INCREASED LENGTH, REDUCED HEIGHT, AND SCULPTURED BODY PANELS ARE FEATURES OF THE 1958 CONTINENTAL.

ROAD TEST OF THE LINCOLN

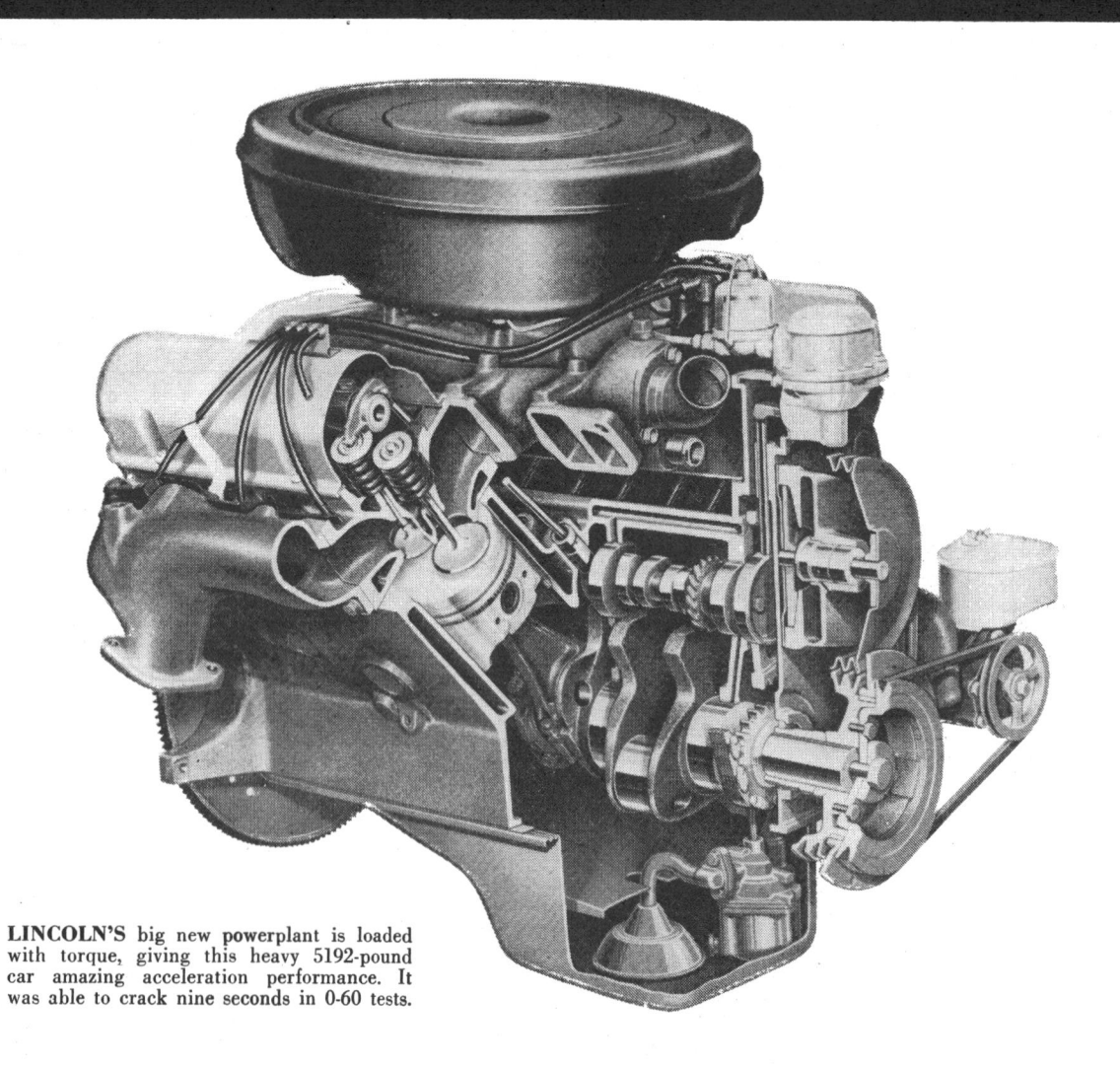

LINCOLN'S big new powerplant is loaded with torque, giving this heavy 5192-pound car amazing acceleration performance. It was able to crack nine seconds in 0-60 tests.

152

Continental Mark III

DESPITE THEIR SIZE, LINCOLNS AND CONTINENTALS ARE MANEUVERABLE, PARKING IS A REAL PROBLEM, HOWEVER.

Lincoln Test Data

Test Car: 1958 Lincoln Continental Mark III
Body Type: two-door hardtop coupe
Basic Price: $5825
Engine: ohv V-8
Carburetion: single four-barrel
Displacement: 430 cubic inches
Bore & Stroke: 4.30 & 3.70
Compression Ratio: 10.5-to-1
Horsepower: 375 @ 4800 rpm
Horsepower per cubic inch: .87
Torque: 490 lb./ft. @ 3100 rpm
Test Weight: 5192 lbs. (without driver)
Weight Distribution: front 2740 lbs., rear 2452 lbs.
Weight-Power Ratio: 13.8 lbs. per horsepower
Transmission: Turboglide automatic (torque converter with planetary gear set, three speeds forward)
Rear Axle Ratio: 2.87-to-1
Steering: 3.3 turns lock-to-lock (20.1-to-1 ratio)
Dimensions: overall length 229 inches, width 80.1, height 56.5, wheelbase 131, tread 61 front and rear
Braking Area: 262 square inches
Suspensions: coil springs at all four wheels (independent front only)
Tires: 9.00 x 14
Speedometer Error: indicated 30, 45 and 60 are actual 30, 42.4 and 55.6 mph, respectively.
Gas Mileage: 10.5 mpg average
Acceleration: 0-30 mph in 3.8 seconds, 0-45 in 5.4 and 0-60 in 9 flat

ALL-NEW is a much overworked term in Detroit, but it definitely applies to 1958 Lincolns and Continentals. Styling, engines, suspension, even method of construction have all been changed.

Most interesting development is the switch to unitized or integral body-frame construction. Lincolns and Continentals are the only Big Three models built in this fashion (although 1958 Thunderbirds are getting similar treatment) and are the largest cars in the world featuring integral construction.

An important reason for the change from conventional body-with-separate-frame construction was that it enabled engineers to cut overall height well below the five-foot mark—which is just what 1957 Lincolns were. And this was done without materially affecting interior room.

Little weight advantage was realized by Lincoln in the switch to unitized construction. In smaller cars, a unitized model customarily is lighter than a similar car with separate chassis and body. In the case of Lincolns and Continentals, however, their sheer bulk required enough reinforcing and stiffening panels to insure adequate rigidity so that any weight reduction was impossible. The new models are heavier and bigger than in 1957.

(Lincolns and Continentals are, in fact, the largest, longest cars now being built in the U.S., except for a few low-volume limousines.)

Despite their size, Lincolns and Continentals are amazingly nimble. The Continental test car amazed all with its excellent performance. It was the first car weighing more than 5,000 lbs. able to crack nine seconds in 0-60 mph tests! Several of these runs were clocked at 8.9 seconds and the overall average was a flat nine.

This is a tribute to Lincoln's new powerplant, a big 430-cubic-inch V-8 loaded with torque. This engine has a number of unique features, including combustion chambers formed in the block, three-stage cooling system and fully water-jacketed intake manifolds.

Top surface of the block on each bank of cylinders is cut at a 10-degree angle, not perpendicular to the bore as is standard practice. Heads are machined flat across, instead of having cast-in chambers. The result is a fully machined, wedge-type combustion chamber—achieved much more economically than would be possible if chambers were cast in heads and then sent through a separate, expensive machining process.

A step cast on each piston crown drives into the narrow part of the wedge-shaped chamber, forcing the fuel mixture across the plug electrode and causing a high degree of turbulence.

These machined chambers ease pre-ignition problems, since carbon and lead deposits do not build up as readily on smooth, even surfaces as they do in rough, cast chambers. In addition, machining insures greater uniformity of volume from cylinder to cylinder. Chamber volume and, thus, compression ratio often varies significantly in cast chambers, making for uneven power impulses and rough-running engines.

A great deal of engineering attention went into proper heat control in these new engines. The three-stage cooling system, for example, was designed to insure fastest possible warmup—important, since a major portion of engine wear occurs during cold starts.

(A welcome side effect was noted. Faster warmup means quicker heat inside the car and more rapid defrosting action.)

Fully-jacketed intake manifolds eliminate the need for conventional manifold "hot spots," were designed to keep the fuel mixture stable and uniform.

Exhaust valves are arranged so that no two are side by side, preventing heat concentrations in localized areas.

Engine breathing is another point emphasized by Lincoln engineers.

Intake valves are 2.15 inches in diameter and exhaust valves are 1.775 inches. (Sizes in the 1957 engine were 2.0 and 1.635, respectively.) Intake manifold design is much cleaner with larger and more smoothly contoured passages. Exhaust passages are larger in cross-sectional area and shorter, allowing less heat transfer from hot exhaust gases to engine coolant.

Despite its increased displacement, this new engine is actually 17 lbs. lighter than the 368-cubic-inch V-8 used in 1957. It is also one inch lower in height.

Attendants at the service station where test cars are fueled and serviced were impressed by the accessibility of most engine components which require regular attention.

Plugs are located above exhaust manifolds and are easy to remove. Distributor and fuel pump are on top and at the front of the engine. The full-flow oil filter can be changed easily from underneath when the car is on a hoist for routine servicing.

The Continental two-door hardtop test car was not equipped with Lincoln's optional air suspension. It had the standard coil springs at all four wheels—which delivered first-rate riding qualities.

Front suspension is much like that Lincoln has been using since pioneering ball-joint front suspensions in 1952. Ball joints are used with long and short transverse arms and coil springs.

Upper ends of the coil springs seat in underbody pockets and shock absorbers are mounted inside the springs. A link-type stabilizer bar one inch in diameter is used.

The rear suspension system consists of two trailing arms, two coil springs and shock absorbers and two compression bumpers. The trailing arms are tied to one of the underbody crossmembers at front and to the rear axle housing at rear. They curve outward and extend underneath the axle, where they are attached to the axle housing through rubber insulators.

The track bar is behind and approximately parallel to the axle. The tubular bar is attached to the left side of the underbody. Rubber bushings are used at all points of attachment between suspension components and the underbody to prevent sound and vibration transmission through the car.

Gabriel shocks are used at front and Munroe shocks at the rear.

A major factor in Lincoln's new suspension layout is that it makes it relatively easy to work in the optional air suspension system. Air springs are simply fitted in place of the coils, the only big change necessary being to move the front shock absorbers back of the rubber air bags.

Other components in the air system are an engine-driven air compressor, air reserve tank, one front and two rear leveling

FULLY-JACKETED INTAKE MANIFOLDS were designed to keep the fuel mixture stable and uniform; they eliminate the need for conventional manifold "hot spots." The design is much cleaner with larger and more smoothly contoured passages. Exhaust passages are larger in cross-sectional area and shorter, allowing less heat transfer from the exhaust gases to the engine coolant fluid.

THE MATING SURFACES of the cylinder heads are machined flat, because of the new in-block combustion chambers in the 1958 engine. In addition to providing a more positive seal between the cylinder block and heads, the flat-surfaced cylinder heads contribute to a more precise control of the compression ratio. Also the machining insures greater uniformity of volume from cylinder to cylinder.

valves, an isolation tank, solenoid valve, check valve and lines and fittings which carry the air to each of the spring units.

With either type of springing, driving torque is transmitted to the body through the trailing arms, which reduces the "wind-up" tendency common to leaf springs. The track bar neutralizes side forces since it ties together the axle housing and underbody behind the rear axle.

One problem long associated with unitized construction has been elimination of noise inside the car. Since the chassis and body are welded together into a single structure, there is often a tendency for road noises and vibrations to be transmitted throughout the automobile. The body, in effect, is turned into a big "sounding box."

Careful attention was paid to this point during the road test and the consensus of opinion is that Lincoln has done an excellent job of noise control. As mentioned above, rubber insulators are used at all suspension attachment points. Various

HEAT FOR CAR INTERIOR is carried through ducts in either front door and directed to rear seat. This setup was also successful on the Mark II Continentals now carried over to the Mark III cars.

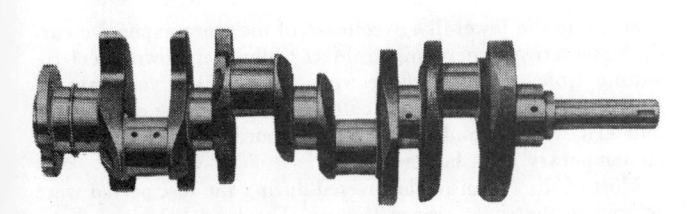

A MORE RIGID CRANKSHAFT is obtained by an increase in journal overlap; it provides reserve strength necessary for severe stresses resulting from higher compression ratio. Also, diameters of the crankpins and main bearing journals have been enlarged.

PROPER HEAT CONTROL in these new engines is a result of careful engineering study. Three-stage cooling system was designed to insure fast warmup. Note extra ducting on aircleaner setup.

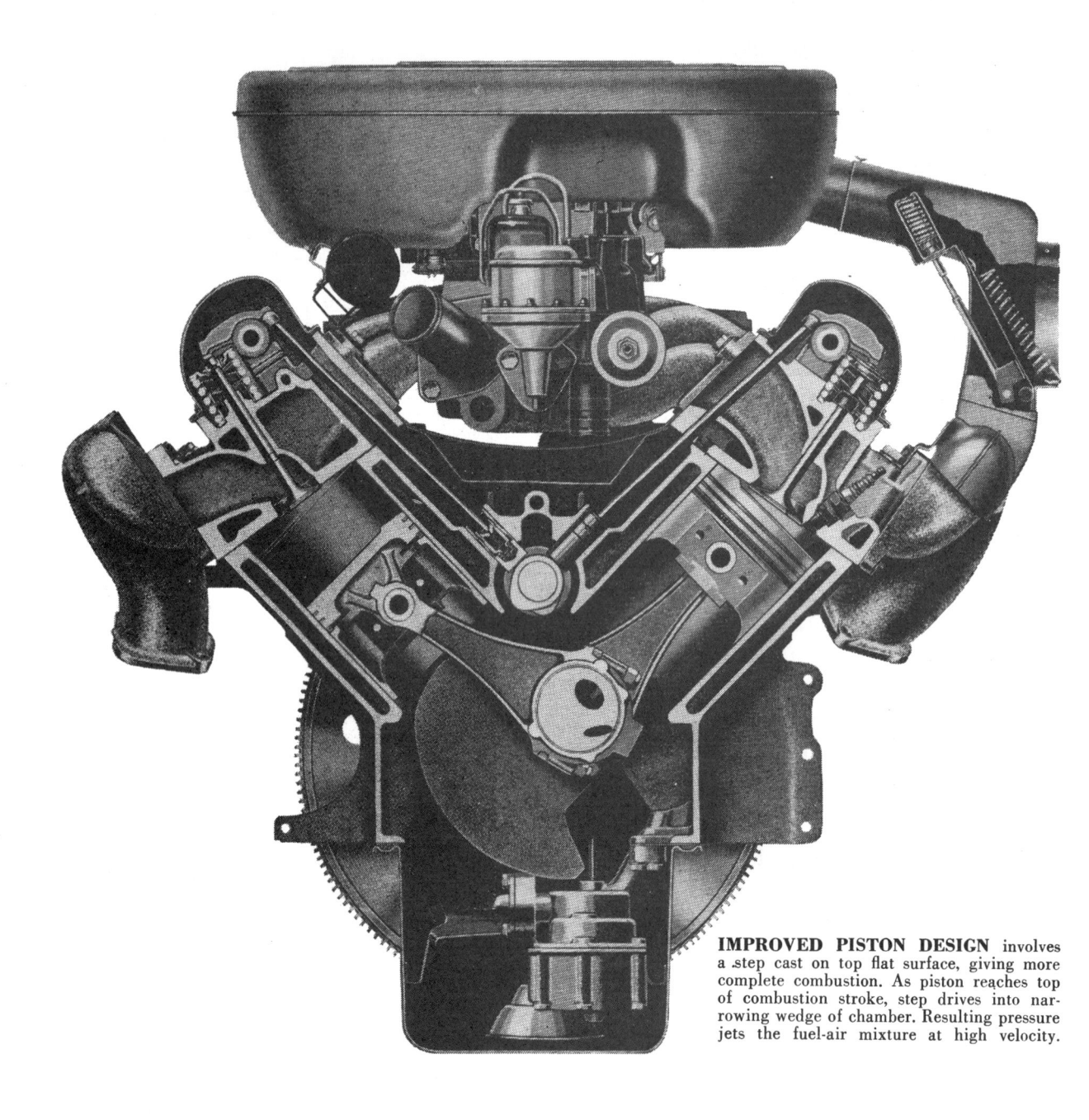

IMPROVED PISTON DESIGN involves a step cast on top flat surface, giving more complete combustion. As piston reaches top of combustion stroke, step drives into narrowing wedge of chamber. Resulting pressure jets the fuel-air mixture at high velocity.

insulation materials have been used extensively at critical areas. The result is the typical quietness and vibration-free comfort associated with luxury automobiles in the Lincoln and Continental price class.

Styling of new Lincolns and Continentals is considerably different from last year. Everything about them emphasizes their impressive size. Straight, horizontal lines run from front to rear, exaggerating the 229-inch overall length.

Major styling differences between Continental models and those in the Capri and Premiere series are at front and rear. Continentals have a more finely-textured, mesh-type grille reminiscent of Mark II cars.

At the rear, Continental series models have three lights set in each side of the grille-like insert above the rear bumper. Capris and Premieres have all rear lights mounted in a single wedge-shaped housing and have a different grillework insert.

Interior of the test car was typically luxurious. Supple leather upholstery was used throughout and the instrument panel obviously was inspired by the Mark II, but does not come up to the jewel-like excellence of the more expensive car.

Conservative owners might object to the step-down effect resulting from depressed floor wells. This means you have to step over the door sills when entering the car. Ease of entrance and exit was not judged to be any more difficult than most contemporary cars, however.

Most of the criticisms uncovered during the test period were related to the sheer size of the car. The long 131-inch wheelbase and 229-inch overall length have their penalties—particularly in parking or other maneuvering in tight places.

The broad, flat hood gives the impression of extreme width, but this is mostly an illusion. The overall width of 80.1 inches actually is two-tenths of an inch less than 1957 Lincolns.

Despite the car's size, it had thoroughly acceptable handling characteristics. Obviously it is not as nimble as the 1952-55 Lincolns of Mexican Road Race fame, but it does have an edge over at least one of its major competitors.

Ride is very comfortable, soft and smooth over practically all surfaces. The long wheelbase eliminates pitching motion,

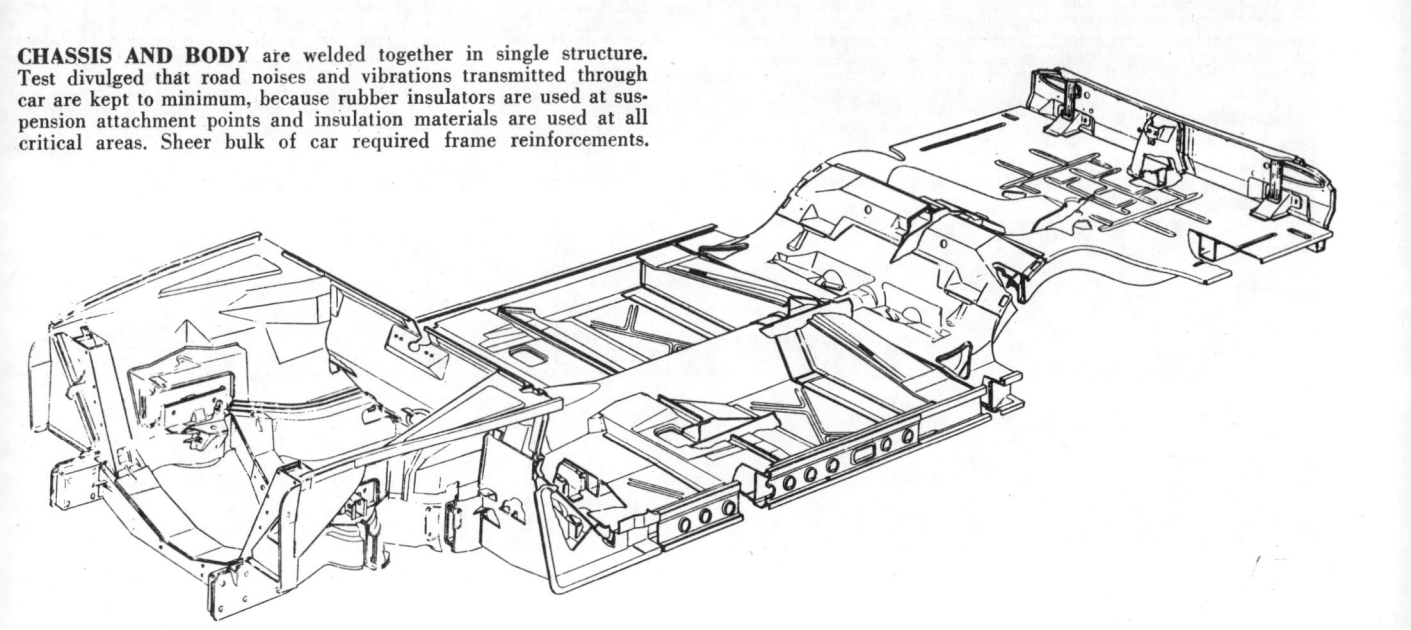

CHASSIS AND BODY are welded together in single structure. Test divulged that road noises and vibrations transmitted through car are kept to minimum, because rubber insulators are used at suspension attachment points and insulation materials are used at all critical areas. Sheer bulk of car required frame reinforcements.

DUAL EXHAUST PIPES run alongside the driveshaft in only "tunnel" in understructure. Coil springs are replaced by air springs at same attachment points in cars equipped with optional air suspension system. Note the clean design of this understructure.

and the over-the-waves effect often associated with very softly sprung automobiles.

Steering is very good. Saginaw recirculating-ball power steering is standard on all Lincoln and Continental models. It is surprisingly fast. Overall steering gear ratio is 20.1-to-1 and only 3.3 turns of the wheel are required to go from lock to lock. It was rather interesting to find that the curb-to-curb turning circle diameter has actually been *reduced* from 1957, in spite of the fact that current models are larger. The turning circle for the test car was 44.4 feet, compared with 45.5 feet for 1957 Lincolns.

Power brakes are also standard on all Lincolns and Continentals. Because of the change from 15-inch to 14-inch tires and wheels on the new models, brake drum diameter was reduced from 12 to 11 inches. To compensate, much wider drums and linings are being used. Both front and rear shoes are now 3.5 inches wide. Width in 1957 was 2.5 inches at front, 2.0 inches rear. This has increased effective lining area from 207.5 to 298 square inches, a 43 per cent boost.

One very fine safety feature discovered in the course of this test was a locking device for the electric power window system. In addition to the usual switches which enable the driver to raise and lower all windows, there is a separate switch that enables him to lock all the individual toggles so it is impossible for passengers to raise or lower their windows.

This is a wonderful precautionary device when children are in the car. It prevents them from playing with the power windows and possibly getting their hands caught accidentally.

Another interesting feature about the windows on the test car was the retractible backlight. Also power operated, this rear window can be lowered slightly for ventilation without creating a noticeable draft. It's fine for clearing the air when you've got a carload of smokers.

The visor effect created by the extension of the roof panel out past the Continental's rear window also proved its worth after the car had been standing outside overnight during a snow storm. The usual vision-impeding blanket of snow was absent next morning! The compound-curved windshield certainly was loaded, however.

Indications so far are that Lincoln has an extremely salable line of automobiles for 1958. Sales were running well ahead of last year at presstime—and a surprisingly large percentage have been of Continental series models. Although the price differential between comparable Premieres and Continentals is roughly $500, Continentals reportedly were accounting for something like 75 per cent of all 1958 Lincoln sales at one point early in the new model season!

Lincoln learned the hard way that luxury car buyers demand impressive size. The compact 1952-55 models never did win the share of the market they deserved. It was this in mind that 1958 models were born—and they have a lot to offer, not only in size, comfort and performance, but also in advanced and rather unique engineering features. •

Given a completely enveloped shape of this size, Lincoln's designers have done well.

ROAD TEST CONTINENTAL MARK III

an astonished inquiry into the shape (and the size) of things at home

THE ADVENT of the huge 1958 Lincoln around Road & Track's offices created more controversy than we've experienced in years. Most of the staff opinions were slight variations on "What could anyone possibly want with that thing?"

But one lone staffer (he shall be nameless) held out: "I like it, its styling is unique and distinctive, it's big and impressive, it's extremely comfortable for long trips, it performs—it is, in fact, a modern Bugatti Royale."

Regardless of whether just everyone wants or could afford a Royale (the largest passenger car ever built; Road & Track, February 1955), the fact remains that the Lincoln is the biggest automobile we have ever driven or tested. Its overall length of 229 inches is such that it will not fit into an 18-foot garage, yet our biggest surprise came when driving this behemoth in traffic.

Steering is, of course, power assisted and very light in action. Full lock from left to right is commendably quick at only 3.8 turns. In traffic one soon becomes used to the sheer bulk of the surroundings and to a slight "lumpy feeling" as the steering wheel is moved off to either side of the true straight-ahead position. In fact, with some practice and some teamwork with a passenger, it is possible to thread the Lincoln through traffic gaps with a great deal of confidence and no heart-rending scrapes. However, visibility forward and over the high hood could be better and would do much to inspire greater confidence. And as usual, there is considerable distortion in the corners of the wrap-around windshield.

The Continental Mark III version of the Lincoln is not appreciably different from the standard Lincolns, but the soft-top body style, as tested, is available only as a Continental. The other three body styles available in the Capri, the Premiere, or the Continental are a four-door sedan, a four-door hardtop and a two-door hardtop. The Continentals have slightly different trim details, including lattice-type grilles front and rear instead of the horizontal louvers used

In sheer amounts of chrome, the Mark III is not far from most American cars. Its great bulk serves to keep the ratio of chrome to painted areas to a fairly modest figure for current trends.

Coming or going, it looks the same and like nothing else.

The mammoth top folds into a metal-covered compartment.

CONTINENTAL *continued*

on the two lower-priced lines. Incidentally, the price quoted ($5792) is f.o.b. Detroit; Federal excise and local taxes, freight, handling charges and preparation must be added.

Driving the Mark III was quite an experience for us. Without a doubt the car understeers, but we didn't try any exuberant cornering, for obvious reasons. On the road the car handles well, rocks and rolls a little too much at high speed over wavy surfaces, rolls moderately in fast turns, but nevertheless is rock-steady and extremely stable on the straight, even at speedometer readings in excess of 120 miles per hour.

During the performance testing we were quite impressed with the initial take-off, particularly in view of a test weight no less than 2.78 tons. Here we can thank the designers for employing an extremely large engine (actually slightly larger than the model J Duesenberg's) and a very low starting ratio of 14.28 (6.80 low gear times 2.1 torque multiplication in the converter). The car literally leaps off the mark with none of that delayed-action effect so common with earlier efforts at adapting torque converters. The net result can be assayed by noting the 0 to 30 mph time of only 3.0 seconds—equalling the best we have ever recorded for an automatic-transmission vehicle.

The automatic transmission has the usual 3 speeds forward and the driver can elect to start in either 1st gear or, for economy, in 2nd. Even 2nd gear starts are fairly brisk, thanks to an overall ratio of 8.92:1. At full throttle the 10.5:1 compression ratio makes some protest via an audible knock, which does not augur too well for the future, when carbon deposits will begin to accumulate. Normal upshifts are smooth and even full-throttle shifts are not too bad, considering the acceleration available. As indicated in the data panel, the 0 to 60 time is only 8.7 seconds, and an honest 100 mph (108 on the speed indicator) requires only 31 seconds to accomplish—rather startling performance by any standards.

Not incidentally, the brakes constricted by 14-inch wheels proved only just adequate for one quick stop from 116 mph. Recovery is fairly quick if a normal cruising speed is resumed immediately.

Technically, the unit frame and body merits considerable praise. The car feels rigid and has, we felt, less shake than any open or convertible-type body we have tried in recent times. Unfortunately, there are some other technical points in this machine which are very difficult to justify.

Consider, for example, the factory-advertised horsepower

In current fashion, the air cleaner hides everything.

Panel retains basic design used for many years past.

PHOTOGRAPHY: POOLE

An attempt at originality in solving a tough problem.

of 375 at 4800 revolutions per minute and the 2.78 axle ratio. As indicated in the data panel, 4800 rpm would be equivalent to 139 mph. If the engine developed that much horsepower as installed, the car might go that fast. The same engine speed with the transmission locked in 2nd gear would give 94 mph, yet the automatic shift point is a mere 3900 rpm and 77 mph (83 mph indicated). Obviously something is wrong; either the engine will not even run at 4800 rpm, or the bhp figure is wildly optimistic, or the gear ratios are very badly chosen. Any design is a compromise, but we feel that the Lincoln has some very strange features in the technical department. Nevertheless, the net result is a car that performs well and smoothly, though it is not particularly economical of fuel. The gear ratios could be altered to give considerably better acceleration and top speed, but this would require a 4th speed to retain the economy they have. Our Tech Ed estimates that the true "as installed" bhp is about 200 at 4000 rpm and may go to a peak of 240 at 4800 rpm. If the axle ratio was 3.31, for example, the top speed would undoubtedly be increased, allowing the full 240-hp potential to be released and giving a top speed of about 121 mph at 4800 rpm. Such a change would also improve acceleration all along the line, but fuel economy would suffer.

The Continental models are finished in conservative good taste, both inside and out. Genuine leather upholstery, specially imported from Scotland, is optional if desired. The folding cloth top of our test car is well designed and beautifully constructed—in fact, it is difficult to recognize the car as a true convertible until one gets within a distance of a few feet. Operation is of course by power, even to electrically driven screws which replace the usual manually operated toggles at the top of the windshield. Two bulges on the rear deck betray the convertible's identity, but appear to be necessary to allow clearance for the retracting mechanism.

Incidentally, what appears to be an enormous trunk space is somewhat misleading. The folded top takes up perhaps ¼ of the volume, with almost another ¼ occupied by the gas tank and spare tire. Nevertheless, there is still enough room in the trunk for an 8-foot dinghy.

Finally, our one staff member who honestly liked the car did admit, at the end, that he would like to see the fender openings enlarged and 8.00-19 wheels and tires fitted. Not a bad idea—next year we could have a Savoy, Premiere, Continental and (maybe) a high-wheeled "Royale."

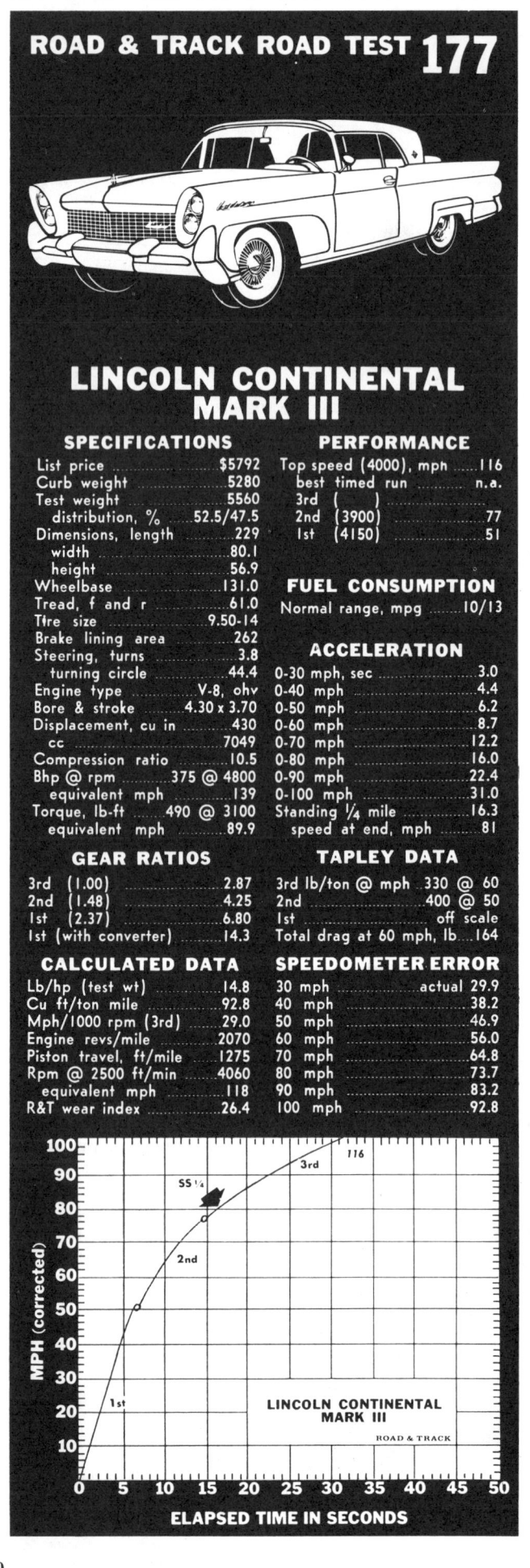

ROAD & TRACK ROAD TEST 177

LINCOLN CONTINENTAL MARK III

SPECIFICATIONS
List price	$5792
Curb weight	5280
Test weight	5560
distribution, %	52.5/47.5
Dimensions, length	229
width	80.1
height	56.9
Wheelbase	131.0
Tread, f and r	61.0
Tire size	9.50-14
Brake lining area	262
Steering, turns	3.8
turning circle	44.4
Engine type	V-8, ohv
Bore & stroke	4.30 x 3.70
Displacement, cu in	430
cc	7049
Compression ratio	10.5
Bhp @ rpm	375 @ 4800
equivalent mph	139
Torque, lb-ft	490 @ 3100
equivalent mph	89.9

GEAR RATIOS
3rd (1.00)	2.87
2nd (1.48)	4.25
1st (2.37)	6.80
1st (with converter)	14.3

CALCULATED DATA
Lb/hp (test wt)	14.8
Cu ft/ton mile	92.8
Mph/1000 rpm (3rd)	29.0
Engine revs/mile	2070
Piston travel, ft/mile	1275
Rpm @ 2500 ft/min	4060
equivalent mph	118
R&T wear index	26.4

PERFORMANCE
Top speed (4000), mph	116
best timed run	n.a.
3rd ( )	
2nd (3900)	77
1st (4150)	51

FUEL CONSUMPTION
Normal range, mpg	10/13

ACCELERATION
0-30 mph, sec	3.0
0-40 mph	4.4
0-50 mph	6.2
0-60 mph	8.7
0-70 mph	12.2
0-80 mph	16.0
0-90 mph	22.4
0-100 mph	31.0
Standing ¼ mile	16.3
speed at end, mph	81

TAPLEY DATA
3rd lb/ton @ mph	330 @ 60
2nd	400 @ 50
1st	off scale
Total drag at 60 mph, lb	164

SPEEDOMETER ERROR
30 mph	actual 29.9
40 mph	38.2
50 mph	46.9
60 mph	56.0
70 mph	64.8
80 mph	73.7
90 mph	83.2
100 mph	92.8

160

LINCOLN rear bumper is molded to become part of body.

CONTINENTAL tail lights have exhaust stack treatment.

LINCOLN FOR '59...

...offers less noticeable sculpturing, minor mechanical refinements, options galore, and another Continental.

IN KEEPING WITH THE POLICY of building outstanding luxury cars with lasting designs, modifications to both Lincoln and Continental models for 1959 have been held at a minimum. The rear ends of both cars look very much like the front, even requiring a "grille" to finish them off. Continental distinguishes itself from the regular Lincoln by exhaust-like housings for stop, tail and back-up lights, and a new center ornament.

Probably the most noticeable change is the lessening of depth in the sculptured scallop in the forward quarter sheetmetal. The Lincoln grille now has narrow horizontal bars that accentuate the lowness and width of the car, while the Continental grille is made of block-type cells in keeping with European trends.

Both cars have lighter-looking wrap-around bumpers, while the fender lines have not changed.

Mechanically, neither car has been changed beyond refinements in engines, running gear, and brakes (most of which have been made from time to time during the model run). They have been confined largely to new engine mounts and redesigned camshafts. Six new rubber mounts have been placed at contact points in the rear suspension to further isolate noises.

A host of optional accessories include these and more: single knob control of air conditioning, special FM radio, six-way power seats, pushbutton chassis lubrication, non-slip differentials, remote control deck lid, tinted glass, and all-leather trim.

CONTINENTAL'S egg-crate grille follows European trend.

SCULPTURED SCALLOP on Lincoln Premiere is shallower.

Lincoln

Imperial

Cadillac

LUXURY AUTOMOBILES ON TRIAL

WHICH LUXURY AUTOMOBILE produced in America today is really the finest and best of all the prestige cars? To find out, MOTOR TREND has made a complete comparative road test of the three top family cars, all four-door hardtops. More expensive models are produced by the manufacturers of the Lincoln, Imperial, and Cadillac, but they are in the limousine or special class, hence do not generally come into consideration for ordinary family use.

Each of the three cars tested is competitive in price, with each manufacturer considering his particular model the most logical choice for the man on the way up. Which fine car should you buy? The choice must inevitably remain with the buyer, but here are a few conclusions drawn after living with each of these cars for more than a week and driving them at a time of year when the worst points of a car manifest themselves more quickly than do the good features.

If you want the widest car, you'll be surprised to learn your choice will be the Imperial. To own the longest, you'll have to make your deal on the Lincoln; it is also the lowest. The quickest away from the light, according to our test, is the Lincoln. The big Cadillac gets the blue ribbon for the best fuel consumption. For the softest ride, Cadillac takes the honors, but for handling and roadability, the Imperial can't be beat.

The brakes on these cars are not too good. The total contact brakes of the Imperial allowed but five slowdowns from 60 to 20 mph before fading (four each for Lincoln and Cadillac) and seven stops before braking effectiveness was totally gone (seven for Lincoln and six for Cadillac).

Steep driveways can cause trouble on all three cars with the long overhang on each the culprit; this can also cause parking problems, of course. The Imperial, especially, suffers from a combination of long front overhang and the lowest *front* angle of approach if dips or some railroad crossings are taken at critical speeds. As the photographs to the left prove, all dip with their nose to the ground over rough crossings.

For the most silent riding car, Cadillac is our choice, but as to wind noise on the open road, there's little choice. Most impressive car of the group is strictly up to you and the car's general reputation in the fine-car field.

What is wrong with these prestige cars? There are faults, many of them inexcusable in such costly machinery. For instance, the Lincoln did not make use of the principal virtue of unitized construction; the entire body possessed many rattles. Driving over rough road surfaces produced pronounced shaking of body panels. Fortunately, these vibrations were not communicated to the passengers, but the noise was objectionable. Doors fitted too tightly; the left rear door closed so tightly that the weatherstripping was mutilated and torn away. Windshield wipers had the poorest sweep and worked with difficulty. The Imperial indicated poor inspection, for the right rear door was incapable of being opened from the outside. The Cadillac had a creaking dashboard which made noises on every bump, and the right front door could not be locked electrically or manually.

Yet, even for these correctible criticisms, all three cars are worthy of being called America's finest automobiles. The one which satisfies the individual whims of the buyer is *his* best buy. They do differ, however, and you'll be better equipped to judge by reading the detailed test report of each car on the following pages.

continued

LINCOLN PREMIERE

ACCELERATION
From Standing Start
0-45 mph 6.2   0-60 mph 10.8

Passing Speeds
30-50 mph 4.3   45-60 mph 4.9
50-80 mph 11.4

FUEL CONSUMPTION
Stop-and-Go Driving
8.8 mpg for 162 miles

Highway Average
12.0 mpg for 186 miles

Overall Average
10.7 mpg for 348 miles

Fuel used: Mobilgas Special

BRAKING
Withstood 4 slowdowns from 60 mph to 20 mph before fading appeared

---

IMPERIAL CROWN

ACCELERATION
From Standing Start
0-45 mph 6.6   0-60 mph 11.9

Passing Speeds
30-50 mph 4.8   45-60 mph 5.5
50-80 mph 11.8

FUEL CONSUMPTION
Stop-and-Go Driving
8.6 mpg for 144 miles

Highway Average
12.7 mpg for 173 miles

Overall Average
10.9 mpg for 317 miles

Fuel used: Mobilgas Special

BRAKING
Withstood 5 slowdowns from 60 mph to 20 mph before fading appeared

---

CADILLAC 60 FLEETWOOD

ACCELERATION
From Standing Start
0-45 mph 6.9   0-60 mph 12.6

Passing Speeds
30-50 mph 4.9   45-60 mph 5.6
50-80 mph 12.2

FUEL CONSUMPTION
Stop-and-Go Driving
9.8 mpg for 109 miles

Highway Average
13.6 mpg for 204 miles

Overall Average
11.6 mpg for 313 miles

Fuel used: Mobilgas Special

BRAKING
Withstood 4 slowdowns from 60 mph to 20 mph before fading appeared

Specifications of above cars on page 71

COMPLETELY NEW, the Lincoln is built in an equally new factory designed especially for the construction of unitized bodies. Thousands of separate welding operations fabricate a combined body and under-frame of great strength and rigidity. The entire unitized body, before final fitting and painting, is dipped in a tank where rust resisting primer flows over and into every nook and cranny.

We tested the Premiere Landau, Lincoln's finest four-door hardtop. It was fitted with a 375-hp engine, Turbo-Drive torque-converter automatic transmission, power steering, and vacuum-assisted power brakes — all of which are standard equipment and are included in the base price. The test car also had a heater and defroster, a radio with an optional station changing floor-mounted step-switch, power windows, six-way power seat — all extra-cost accessories.

In the Lincoln, you step down when entering, unlike the other two luxury automobiles tested, making the low roof less of a hazard. Elderly persons especially will find the Lincoln rear seat the easiest entered of the three cars, largely due to the lowered floor. Instruments and controls are the best located of the three cars. The driver need never stretch as all controls for heater, fresh-air ventilation, radio, lights,

MT'S TEST IMPERIAL was the luxurious Crown Southampton four-door hardtop. Standard equipment included the TorqueFlite automatic three-speed transmission, power steering, and power brakes; the extra-cost accessories installed on the test car included the radio with floor station-switch, heater, air-conditioning system, six-way power seat, and power windows.

Since the floor was level, not recessed as in the Lincoln, we found ourselves ducking our heads slightly to avoid knocking against the top sill when entering the car. Rear seat entrance is made less difficult because of the unusually wide door opening.

Instruments and controls run a close second to the Lincoln with one exception — there's little logic in the clumsy placement of the turn signal switch at the bottom end of the vertically-arranged transmission pushbuttons. This requires one to remove the left hand from the steering wheel, glance quickly, and reach about nine inches down and to the left. Window controls are in the door panels, and the electric door-lock switch (optional) is positioned for easy left-hand operation. Instrument lighting is the best of the three cars — the black lighting does not glare, makes reading the large dials

THE BIG, FLEETWOOD 60 SPECIAL was MT's test Cadillac. Available only as a four-door hardtop, this car has 3.5 inches more wheelbase and a scant two inches more overall length than the variety of models available in the more popular 62 series. The 60 has greater rear legroom than the smaller cars, but the same space in front. The test car was equipped with radio and heater, air-conditioning, a six-way power seat, power windows (including the vents), electric door locks, and air suspension. Its engine was the 310-hp unit equipped with the standard single four-barrel carburetor. An optional "Q" engine uses an improved intake manifold mounting three two-barrels, raising the hp to 335, with the same rpm torque.

Entry and exit are on a par with the other large cars tested, but the front seat cushions are on the short side, giving less leg and thigh support than on the other cars. Power seat controls are in the driver's armrest, making adjustment easiest of the three cars. Floors are level with the door sill, making entry into the rear more difficult due to the downward sweep of the roofline at this point.

The electric door locking controls are impossibly positioned far back on the door sills, requiring difficult hand etc. are neatly gathered into the bottom half of the instrument group which occupies a separate panel directly in front of the steering wheel. The instruments, while excellently placed, reflect badly in the upper left half of the windshield at night; however, only half this glare is the fault of the panel lighting — the real culprit is the prism-like transmission quadrant.

Driving position is excellent and the seats are deep enough to give genuinely comfortable support. In traffic, the steering is quickly responsive, although we found the test car cruised with better than one-half turn of free play in the steering wheel over the specified 3.3 turns lock-to-lock. One is amazed at the maneuverability of this huge car. We found that setting the transmission lever to "L" (low) would drop the speed gently to second gear without jerking; when the speed is dropped to about 20 mph, first gear comes in with greater engine braking, but very smoothly. The more you drive this big, heavy car you become aware that it does not ride as easily as you had expected. Although bumps are felt only slightly, you feel them more than in the other two cars tested. Roadability is almost superb — directional stability is the best I have ever experienced in any car, small or large.

Corners at speed are easy, the car's trailing arm rear suspension giving just the right amount of understeer. Although there is considerable lean on sharp or fast corners, this is not too apparent to drivers or passengers. Such heeling-over is less than that of the Cadillac, more than the Imperial.

Luggage space on the Lincoln is cavernous, the most usable capacity of any of the three cars tested. Once warmed up, the heating and defrosting system will run you out of the place. Especially good is Lincoln's method of piping heat to the rear passengers. The defrosters were the best of the three and covered the greatest expanse of windshield.

The Lincoln has a feel of luxury; its leather upholstery is carefully fitted over seats that were the softest of the three cars tested, and folding armrests in the centers of both front and rear seats add greatly to highway comfort. The rear center armrest, however, should have a pull tab of some sort to facilitate unfolding it from its very tight fit in the backrest. The rolls and pleats of the upholstery have a custom look and feel, and the carpeting is thick and soft. Separate lamps at each extreme end of the dash light, with the dome light, when either door is opened, giving this car the best interior lighting of the three. /MT

---

easy, and there is almost no annoying reflection at night in the windshield.

The steering wheel, in our opinion, is the best of the three; the straight-across spoking allows a relaxing change of hand positions on a long trip. The power-seat controls are positioned handily at the bottom left of the seat; with the seat in the far aft position, there's a disturbing space of uncovered steel floor plainly visible beneath the front edge of the seat. Newly positioned left of center, the rear-view mirror does give improved view aft, if there is no-one sitting in the middle of the rear seat. This mirror, along with several others checked, failed to stay put after adjustment.

Driving position ranks, in my opinion, the best of the three cars tested. The depth of the seats from front edge to the backrest gives fine leg and thigh support. Armrests are comfortable; the front rests feature a hinged top surface which discloses a roomy space for maps and other small items like flashlights, fuel coupons, etc.

In driving, we often had the feeling, when catching sight of either fin in the corner of our eye, that another driver was too close. Fast 3.5-turn steering makes traffic maneuverability good; the lack of any free play in the steering made this car the quickest to respond of the three. The Imperial, with its torsion-aire suspension system, handles like a well-trained quarter horse on all road surfaces. On the test car were the optional 11.00 x 14 tires which give a softer ride. Unfortunately, however, these fatter tires, though they increase traction, do not help roadability. On the contrary, they sometimes cause a slight rear-end sway on quick turns. Power steering is quick and responsive; there is slight understeer, just enough to facilitate high-speed travel in greater safety. Cornering lean is slight, less than the other two cars. We noticed a hint of road wander, brought on most likely by the large tires.

Heating and ventilation are good, but defrosting is the least effective of the three cars tested. Heat, however, comes within three minutes after a cold start.

The feel of luxury in the Imperial is emphasized by the metallic-threaded nylon upholstery; it looks and feels rich. Warmer than the leather in the Lincoln in cold weather, this upholstery is matched by the finest looking door panelling, the best formed and most comfortably located center armrests. Chromeplating on the instrument panel has a better look of quality than on the other two cars. /MT

---

maneuvering. The outside rear-view mirror is controlled by a handle from the inside. The parking brake gave trouble on freezing mornings; the release trip is on the left of the pedal and sometimes release is not positive. Instruments are surrounded by too much chrome; at night, unless the panel light is dimmed well, reflections in the windshield are bothersome. Foot pedals for throttle and power brakes are close enough for simple foot pivoting.

Since the Cadillac had air suspension, we used the height control lever to raise the car to climb a curb when it was pinned into a parking slot by a double-parked car. It takes two or three minutes to raise the car; the tail end rises first, followed more slowly by the front end. Road clearance can be increased by more than four inches for emergencies.

The steering wheel has two spokes with nearly the shape of a knife edge, not making them a safe resting place for the driver's chest in case of an accident. As in the Lincoln, we favored the automatic transmission control lever over the dash-mounted pushbuttons that are on the Imperial; you do not have to remove either hand from the wheel to operate this or the turn signal switch.

Driving position is good, although the driver is the least comfortable of the three because of the short seat design. Front armrests on the doors are not positioned for best comfort; a recess for pulling the doors closed is located exactly where the elbow rests. The transmission selector indicator is centered immediately below the horizontal speedometer face, best for viewing of the three. In traffic, only disturbing factor is the slow steering which requires considerable winding, nor does the steering wheel effectively center itself after a sharp turn as on the other two cars tested. Downshifts for safer braking are smooth. Visibility to the rear is best from behind the wheel of the Cadillac.

On trips, the air suspension gives a beautiful ride, but road bumps, dips and paving strips are still felt. The Cad has greater lean on corners than the other two; nose diving on an emergency stop is greater, also. Road feel through the steering wheel is less, however.

Heat circulation is good; there were no leaks around doors or windows, and defrosting of the windshield was excellent.

As to the feel of luxury, the test car had beautiful nylon upholstery in a dull gray tone to lend subtle charm to the interior. Rear passengers now have manually-operated vent panes in the windows. /MT

LINCOLN shares a facelift and technical changes with CONTINENTAL

FORD'S two big luxury liners, the Lincoln and Continental, haven't changed personalities in the switch from '58 to '59. But they have been given enough of a facelift and heady engineering improvements to make them a continuing big factor in the high-priced field. For any money Lincoln remains one of the largest packages on the road today.

The grille texture is continued into the headlamp doors in 1959 to achieve a wider looking front appearance. Bumper lines have been smoothed by the removal of the bullet-shaped bumper guards. The horizontal lines of the bumpers, upturned at the ends, follow the contour lines of the grilles.

Aside from a depression for mounting front license plates, bumper smoothness is unbroken across the full width. Parking lamps in 1959 are located in circular, recessed pods at the extreme ends of the bumpers. For an added touch to the frontal appearance, stylists have included chrome lip moldings. The moldings extend across the hood and blend into the chrome around the headlamp doors.

The long look has been further accented in the '59 models by the front fender scallop which has been extended rearward into the front door area. The extension of the scallop and the addition of front bumper extensions assist in maintaining the appearance of length. The bumper extension is mounted in the fender scallop area ahead of the wheelhouse.

A chrome spear molding extends from the scallop to blend into the rear chrome treatment. A brushed-finish applique is applied to the lower rear quarter panel and fender skirt. This is made of brushed aluminum. Chrome rocker panel moldings and dual, rear-mounted antennas are available as accessory items.

The Continental series is identified in a side view by the rearward slanting rear window frame. The chrome frame comes down on a slightly angled vertical line and extends sharply rearward along the upper back panel to the rear deck line. The "Continental" name is mounted in the forward section of the front fender.

New wheel covers are styled in two designs—one for the Continental series and the other for the Lincoln and Premiere series. The Continental covers have a satin finish, chrome hubs with chrome spokes extending out to the rim. The areas between the spokes are painted black.

The 1959 Lincoln and Premiere series running and stop lights are combined into one integral unit—a horizontal oval that follows the contour of the rear grille oval. Back-up lights are located inboard of the taillights and are built into the rear grille.

The rear grille texture consists of horizontal chrome strips extending across the lower back panel.

Most striking feature of Continental's rear styling is the flat, reverse-sloped, retractable back window which is framed by chrome window channels. This feature is also carried into the convertible model and eliminates the plastic back window used throughout the industry. Hardtops have a chrome fluted hood projecting out beyond the back window for protection against snow and sun.

Continental's rear grille has three rectangular, chrome-bezeled lights on each side. These are the stop light, running light and the back-up light, in that order from the outboard edges inward. The rear grille is a lattice pattern of chrome with black squares.

The 1959 Lincoln Continental engine is essentially the same as its '58 powerplant. But there are some improvements. Increased smoothness of operation is assisted by new engine mounts, a redesigned camshaft, a new spark curve for the distributor, revised carburetor calibrations, and a new air cleaner.

Both Lincoln and Continental use the same engine. It's a 90-degree V-8 ohv with a piston displacement of 430 cubic inches, a bore of 4.30 and a stroke of 3.70 inches.

Cooling with the unique three-stage cooling system is improved due to the use of a new high-temperature 180-degree thermostat in the manifold. A 160-degree

LINCOLN AND PREMIERE series feature a zinc die-cast grille which gives the appearance of stacked rectangles. The rectangular appearance is sensed in only direct front view. Each segment is peaked slightly to form a broad-based isosceles triangle when seen from side.

THE CONTINENTAL is recognized easily by it's lattice work grille carrying its gold-script name. New this year is the flow of the grille into the headlight section. Lamp mounting remains same.

RUNNING AND STOP lights of the Lincoln and Premiere series are combined into one integral unit. Back-up lights are located inboard of the taillights and are built into the horizontal theme grille.

THE LONG LOOK IS FURTHER ACCENTUATED BY THE TRAVEL OF THE FRONT FENDER SCALLOP INTO THE DOOR AREA AND BUMPER EXTENSIONS.

thermostat was used in early 1958. The new thermostat will allow a higher build-up of manifold temperatures and a consequent reduction in use of the automatic choke for fuel saving.

A new Twin-Range Turbo Drive transmission, which provides for an automatic low-gear start and considerably reduced converter slippage, is available on the 1959 models. Unlike the transmission used in early 1958, this new transmission provides for a fully automatic and modulated shift between first and second gear.

Both cars will again make use of a unitized body as in 1958, except that the '59 structure is about 55 pounds lighter. The unitized body, in which the body and frame are constructed as a single unit, is a complete departure from the conventional, bolted-together body and frame units.

Ford has also developed a rust-proofing process for use on the unitized body. Replacing the spray methods used on conventional cars, the 1959 Lincoln and Continental bodies are protected against rust by immersing the entire body, to the belt line, in a paint tank of special rust-resistant paint. This "bath" technique provides protection against rust to internal and external surfaces of the underbody assembly including side rails, box members, and crossmembers, company engineers say.

Sound insulation is assisted by over 100 pounds of insulating material in each car.

The sound barrier, used as a dash insulator, previously consisted of a layer of deadener material, backed by a layer of fiberglass and followed by a layer of fiberboard. This dash insulator is reported to be vastly improved for the '59 cars by the addition of another layer of aluminum, after the fiberglass, to form an effective "sandwich" of materials. The improved dash insulator, engineers say, provides greater suppression of engine noise under acceleration or wide open throttle operation.

As in the last three years, the 1959 Lincoln and Continental will offer hydraulic power brakes as standard equipment. The wide brakes of the Lincoln have 262 square inches of brake lining area, maintaining the 44-per cent increase gained in 1958 over the '57 Lincoln.

A featured addition to the '59 brake system is the inclusion of an automatic brake adjuster. Available as standard equipment, the adjuster is engineered to each individual brake.

This feature is designed to eliminate brake adjustments by maintaining efficient clearance between the drum and brake lining. The automatic adjuster retains a clearance of approximately 10 to 15 thousandths of an inch at all times. The adjustment is made when brakes are applied as the automobile moves in reverse.

Dealers will have plenty of new items to bring to the attention of their customers with the new 1959 Lincolns and Continentals. •

CAR LIFE 1959 CONSUMER ANALYSIS

LINCOLN

By JIM WHIPPLE

THE '59 LINCOLN is almost a carbon copy of the '58 model. Appearance changes are slight — the major one being the treatment of grille, front bumper and fender and the molding around the canted headlights. The general effect of the '59 front-end facelift is to cool off somewhat the "angular" look of the 58's lights, grille and bumpers.

The body interiors remain unchanged dimensionally, with wide seats, good headroom and adequate height above seat cushions front and rear. Rear seat headroom of 33.8 inches is found in both the sedans and two-door hardtops. Rear compartment leg-room is the best by three to four inches of the cars in Lincoln's class. This dimension of 46 inches holds for the two-door hardtops as well as the four-door sedan and hardtop.

Seat cushions are higher by several inches on Lincoln. Front cushion is 1.5 inches higher (two inches on two-door hardtop), measuring 11.5 vs 10 inches. In the rear compartment, Lincoln seat height on all models is 14.5 inches. On some of the competing sedans (Cadillac), the rear seat comes close with a height of 14.2 inches, but on most other four-door hardtops the rear seats are 12 or 13 inches and 11.7 on the two-door hardtops and convertibles.

What these figures add up to is that Lincoln—with an overall height of 58 inches (curb weight—no load)—has more vertical room in the passenger compartment and provides more conventional, chair-height seating positions.

Although the door openings are no taller—seat cushion to top of open door frame—the fact that passengers are sitting higher off the ground in the Lincoln makes entrance and exit a bit easier than in the other two super-low cars, Cadillac and Imperial.

EXTRA ROOM

The reason that we've put these dimensions and facts at the beginning of our report is simple. The Lincoln's dimensions—interior height and seat cushion height—represent the most important differences between Lincoln and its competing cars. In this roomier interior design Lincoln offers an exclusive advantage that might very well be the deciding factor for comfort-conscious upper-bracket motorists.

For travellers it's well worth noting that Lincoln's high, square rear deck provides more useful trunk space than is found under the tapering rear decks of either Cadillac or Imperial.

After a few hundred miles out on the road with the Lincoln sedan we found two surprises in the car's overall performance.

First of all, the unit-body-and-frame construction—although it banishes all rattles, squeaks or creaks—is not totally immune from shake. The amount of shake was slight and noticeable only on the roughest pavement under medium-to-high speeds.

It was approximately equal to the amount of shake we found in the conventional separate body and frame construction of our test Cadillac. In neither car was this shake ever annoying, or even in any way unpleasant.

STRANGE PROBLEM

It is known in the automotive industry—and among a small number of unhappy owners—that some of the

LINCOLN is the car for you

if... You want the roomiest and easiest to enter of all 1959 luxury cars.

if... You want a big, massive, roomy car that's surprisingly easy to handle and remarkably good at holding the road.

if... A well-balanced package of styling, comfort, roadability, workmanship and performance makes sense in your luxury car choice.

LINCOLN SPECIFICATIONS

ENGINE	V-8
Bore and stroke	4.30 in. x 3.70 in.
Displacement	430 cu. in.
Compression ratio	10.0 to 1
Max. brake horsepower	350 @ 4400 rpm
Max. torque	490 @ 2800 rpm
DIMENSIONS	
Wheelbase	131 in.
Overall length	227.1 in.
Overall width	80.1 in.
Overall height	56.7 in.

TRANSMISSIONS
Turbodrive.

Mark IV limousine has a padded, landau-type roof, formal rear window, is fully equipped and available in black only.

Jim Whipple finds comfortable seating inside 19 feet of Lincoln Capri hardtop. Unitized construction, extensive tests to eliminate resonance points make for very quiet car.

Lincoln has cleaned up front end for '59 with a straight, simple bumper bar. Immense rear deck of Continental slopes to a honeycomb "grille" which is unchanged from '58 except that taillights have been grouped separately.

'58 Lincolns had unpleasant amounts of internal vibration. This seemingly was caused by factors inherent in the large structure of welded steel stampings that make up the unit body-frame construction.

The whole body or hull of some Lincolns behaved like a giant tuning fork when certain vibrations from wheels or engine and driveline hit. Lincoln engineers worked frantically to correct this problem when it showed up in existing cars and to eliminate it from all succeeding production.

Evidently this problem—which was Gremlin-like in that factors normally known to cause body vibration were not involved—did not occur in any prototypes of the '58 model before Lincoln went into production. No amount of normal quality control could seem to eliminate the random occurrence of a "vibrant" Lincoln for which the only cure seemed to be an agonizing amount of down time for modification in the field.

The 1959 Lincoln that I tested was selected from dealer stock and was as good a random sample as you can get. There was not a trace of internal vibration in the car no matter what the speed or road conditions. It didn't seem as rigid as the '58 (see above). Could be that building the body a little less rigid eliminated the vibration problem.

The second surprise was the really excellent handling and roadability of the car. I tested the big Lincoln on the sports car race track at Lime Rock, Conn. I've driven some first-class sports cars on this track, so I had a yardstick of handling performance with which to measure the Lincoln. I'd be a good deal less than honest if I claimed that the Lincoln came near to sports car performance.

But, with 30 lbs. of air in the tires it proved to be as stable and easy to handle as several lower-slung hardtops which are usually praised as far superior "road cars". In all-out cornering, the big Lincoln leaned a bit more than a Cadillac or Imperial would in similar conditions, but the rear wheels stayed put every time and the understeer effect on the front wheels was predictable and never permitted the steering and handling to become mushy.

What it boils down to is simply that the Lincoln is a sure-footed road car, stable and easy-handling, that will eat up mile after mile of narrow, winding beat-up concrete roads at 50 and 60 mph without tiring the driver or tossing the passengers around.

NEW TREND

The '59 Lincoln is a good example of the new trend in the luxury car field toward cars that are so light on their wheels that they seem to deny their locomotive dimensions. I proved this by following a good driver in a popular and extremely capable sports car down a narrow 1½-lane macadam byway.

The big Lincoln could have carried my friend's sports car in the back seat and trunk, but no matter how the road snaked in front of us, the Lincoln kept up with him.

Lincoln goes about its business smoothly and quietly. The short stroke, 430 cubic-inch engine develops 350 horsepower at 4400 rpm and delivers a brutally powerful 490 lbs. of torque at 2800 rpm. The rear axle ratio is a sensible 2.89 to 1, which means that at normal highway cruising speeds the Lincoln paddles along with no sound and fury from beneath its massive hood.

When you do need action, a quick move of the transmission control or

Above, Jim Whipple checks on the accessibility of the 350 horse Lincoln V-8, and finds the engine compartment packed but well laid out.

a flooring of the accelerator downshifts to intermediate (2nd gear), and the big V-8 delivers maximum torque to the rear wheels—right now! Under full-bore acceleration from a standstill, the Lincoln jumped to a corrected 60 mph in just 9.5 seconds, which is lots of jumping for a car weighing 4986 lbs. dry.

Gasoline mileage in high speed cruising and test driving was just a fraction over 10 miles per gallon. This is not economy stuff, considering that the 10.0 to 1 compression ratio of the engine demands premium fuel. But, as we've said in the past, anyone who's alarmed at 10-miles-per-gallon fuel consumption shouldn't be buying a $5,000 car in the first place.

SUMMING UP

Lincoln is the biggest of the big cars insofar as useable passenger and luggage space is concerned. It's also an extremely able, easy-handling road car, very pleasant to drive. It is quiet and, if our test car was a good example, free from annoying vibration. Seating is exceptionally comfortable, vision excellent on all sides and workmanship generally satisfactory.

LINCOLN CHECK LIST
5 CHECKS MEAN TOP RATING IN ITS PRICE CLASS

Category	Description	Rating
PERFORMANCE	Lincoln is the match for any of its competitors in acceleration and will shade many smaller, lighter cars.	✓✓✓✓
STYLING	Lincoln's styling is massive but unusually well-proportioned and happily free from indiscriminate splashes of chrome.	✓✓✓✓
RIDING COMFORT	Lincoln comes close to the top of its class in a three-way contest that's closer this year than in a long time. All three big cars are closer to perfection than they've ever been.	✓✓✓✓
ROADABILITY	Lincoln measures up to its competitors in this category and that's very good going indeed. It comes within a small margin of being the very best handling of them all.	✓✓✓✓✓
INTERIOR DESIGN	Here Lincoln's superior rear seat legroom and higher seating make for a more commodious and easier-to-enter interior than either of the competing cars. Lincoln rates a clear first place in this category.	✓✓✓✓✓
EASE OF CONTROL	Precise power steering with good road feel plus good power brakes and flexible automatic transmission make Lincoln a very-easy-to-handle car in spite of its size.	✓✓✓✓
ECONOMY	Lincoln's big 430 cubic inch engine has a healthy appetite for premium gasoline when you're in a hurry. Easy-going drivers can expect to do better than the 10.5 miles per gallon logged on test run.	✓✓✓
SERVICEABILITY	More reachable engine components make routine service operations such as tune-ups fairly easy.	✓✓✓
WORKMANSHIP	Lincoln workmanship and finish are very good, but not up to the leader in the luxury field.	✓✓✓✓
VALUE PER DOLLAR	Lincoln is a comfortable, roomy, powerful and very roadable automobile. Depreciation is fairly high but the car's reputation is building up and trade-in value should improve.	✓✓✓✓

LINCOLN OVERALL RATING... 4.0 CHECKS

A PIONEER in the modern era of unit-body construction, Lincoln stays with the basic package which was unveiled in 1958. The Continental Mark V varies only by slight trim changes and a mesh-like design of the grille. All-new for both cars is the bumper which features built-in parking lights. Under the hood it's interesting to note a lowering of horsepower and torque ratings.

LINCOLN

there's little change in the luxury car from Dearborn

TWO words, "big and luxurious," describe the 1960 Lincoln and its handsome stablemate, the Continental Mark V.

The first adjective applies to the styling which, except for a few minor alterations, is a direct carryover from the square-line school of design that Lincoln initiated in 1958.

The second qualifier stands for the more resplendent array of paint and trim combinations that are available to adorn this year's car. But a new estimate of the Lincoln must be added to these two when one experiences the way the new version of Dearborn's high priced vehicle responds under power!

This year Lincoln builders had to forego any major sheet-metal changes and decided to concentrate their efforts on improving the ride, road temperament and reliability of the car.

The most significant changes took place underneath the car with the rear suspension the foremost improvement. Now, instead of "dipping" and "diving" when power is applied or released at the rear wheels, the car's take-off and braking behavior is quite normal. Reason? A new Hotchkiss-type rear spring setup that incorporates asymmetrically mounted, longitudinal leaf springs, has replaced the coil arrangement of earlier Lincolns.

This new system (similar to the 1959 Thunderbird, only beefier) also uses re-positioned upper and lower shock absorber mountings that contribute a great deal to an improved roll stability and makes for superior cornering characteristics.

The self-adjusting power brakes (standard equipment on both the Lincoln and Continental) also have been improved with new, thicker lining fitted, upping the lining life by more than 60 per cent.

In the styling area, the basic lines of the 1959 car remain for another year. The frontal treatment is new with a repositioned headlight arrangement and a new bumper line. The rear "grille" on both the Lincoln and the Continental is new as is the back bumper.

A new roof line is found on the Lincoln models with an increased glass area at the rear. Aside from these few minor alterations though, the car exhibits the same large, streamlined box look that has been the vehicle's hallmark since the current styling was used on the 1958 Lincoln and Mark III Continental.

Superior finish and attention to quality control seems to be evident on and in the new car, a very important factor when a unit carries an over-$6,000 price tag.

Inside, the new Lincoln exhibits quiet taste with a restyled instrument grouping and a new dash panel. The instruments are now found in four pods across the driver's portion of the panel with the various light and heater controls underneath.

Under the large, flat hood, new economy and longer life is the theme. A completely new two-barrel carburetor replaces the familiar four-barrel unit of 1959. This new carburetor is said to improve the engine's idle, increase the mpg ratio and provide a better air-fuel mixture distribution.

It's interesting to note here that the fitting of a two-barrel carburetor has lowered the Lincoln engine's horsepower and torque rating significantly. In 1959, the figures quoted were 350 and 490 for horsepower and torque respectively. This year, with the smaller and more economical carburetor, the figures now read 315 and 465 respectively.

Coupled to this more economical powerplant is a refined version of the Twin Range Turbo Drive transmission that has been a Lincoln standard. The unit has been "tuned" according to the car's engineers to deliver increased acceleration, smoother performance and more economy. The latter benefit could only be substantiated by a thorough road test which the editors of MOTOR LIFE will conduct in the near future. But, as for the smoother, more rapid acceleration claims, one brief ride in a new 1960 Lincoln would convince even the most skeptical. This car has to be driven to realize the major change in its handling and performance personality. •

BIGGEST IMPROVEMENT for 1960 is in the rear suspension department. Gone are last year's coils which led to "dipping" and "diving" and are replaced by Hotchkiss-type longitudinal leaf springs.

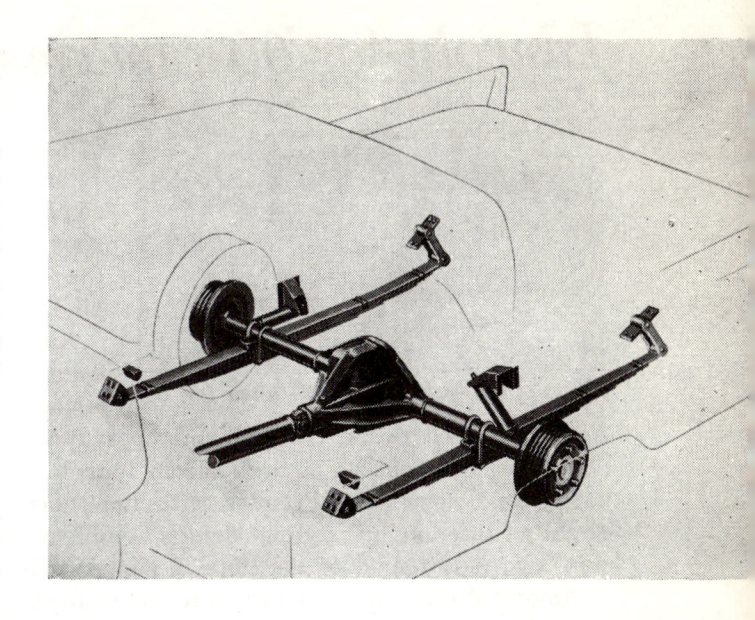

REAR BUMPER features new, straight lines. On Lincoln models the glass area at the rear has been increased by an alteration of the basic roof line. Visibility is somewhat improved by the change.

NEW INSTRUMENT grouping reverts to the familiar aircraft or cockpit theme. Printed circuits are employed in the wiring of the instruments carried in the four large pods across the top.

1960 ANALYSIS
LINCOLN

ENGINES

Cubic Inches	Type	Compression	Carburetors	Torque	Horsepower
430	ohv V-8	10.0	2-barrel	465	315

NOTE—The same engine is fitted in all Lincoln series cars.

DIMENSIONS COMPARED (in inches)

Car	Wheelbase	Length	Height	Width	Front Tread	Rear Tread	
Lincoln	131.0	227.2	56.7	80.3	61.0	61.0	1959
Lincoln	131.1	227.1	56.7	80.1	61.0	61.0	1960

NOTE—Above dimensions are for 1960 Lincoln, Lincoln Premiere and Continental Mark V four-door sedans.

Lincoln for '60 remains long and elegant

430 cu. in. V-8 has been de-tuned for '60.

FOLKS who like their living rooms on wheels and who crave easy-chair, carpet-slipper comfort in 80 mph cross country travel could go a long way before they found the equal of the '60 Lincoln, a true Leviathan of the turnpike.

Although it has basically the same unit body frame as last year, and therefore the same interior dimensions, several significant changes have improved riding comfort and operational efficiency.

The '60 line consists of the Lincoln four-door sedan plus two- and four-door hardtops. Next comes the Premier with the same three models. The Continental series, slightly different in styling details, emblems and trim, is made up of a four-door sedan, two- and four-door hardtops, a convertible, a limousine and a town car.

Among the more noticeable exterior changes are a new rear roofline and back lite, a much smoother and sleeker looking bubble-type window and a smaller, more conventionally shaped front bumper with rubber cushioned guards to fend off the hardware of other vehicles. However, the Lincolns look much the same as they did last year—long, low, squarish and massive. The new rear roofline has improved the overall appearance a great deal though, smoothing out the lines of the car.

Inside, I found that in Lincoln this year as last, your luxury car dollar buys more sheer space than in any other car. The rear compartment is especially noteworthy for the tremendous stretchout room. Width is, of course, well over 60 inches front and rear, and the rear seat is high enough over the transmission tunnel to make three adults unusually comfortable. The cushioning has no peer in U.S. mass produced cars, with personal preference being the guide in choice of upholstery.

New dash is well laid out, but too much chrome makes it hard to read at night.

CAR LIFE CONSUMER ANALYSIS

LINCOLN

The major mechanical change made on the Lincoln this year is a new rear suspension setup, a return to the leaf spring Hotchkiss drive used by Lincoln before the new unit construction body and chassis. The rear axle is mounted forward of center on the springs, so that spring wrap and rear end squat is reduced on acceleration.

I found that although the '60 Lincoln ride is approximately the same as the '59 model on uneven roads, there is a great improvement in ride over rough surfaces such as broken concrete, cobblestones, rough gravel etc. There is an almost total absence of harshness from such surfaces due to liberal use of rubber insulators between the axle and the spring.

Lincoln engineers have really gone to war against vibration this year. There is a new rear engine support consisting of five synthetic rubber cushions on an extra cross member. A new cast iron transmission housing incorporates a hydraulic damper assembly between it and the chassis to damp out engine and driveline vibration. As a result of this extensive re-engineering of engine mounting and rear suspension the big Lincoln is really vibration free. Coupling these improvements with the now rock solid unit body and frame, which is loaded with sound deadening material and insulation of one kind or another, makes the Lincoln move with velvety ease.

The super smooth ride hasn't been achieved at any great sacrifice of roadability. Lincoln has a low center of gravity and wide tread plus a stabilizer bar and shock absorbers that do a great job in cutting down on roll, sway and swerve in cornering. The most amazing thing to me is the easy, light handling of the '60 Lincoln. It doesn't feel like a big car and believe me it is a big job. Yet, aside from the obvious problems in maneuvering 227 inches in limited spaces, you never get the feeling that the car is slow or ponderous in response.

You can whip the car quickly from one lane to the next at 50 mph as readily as a light compact car. There is no feeling of roll, or a surge of uncontrolled weight as sometimes occurs in cars of Lincoln's bulk.

Most of the secret is in the stable, sway-free suspension and the well-designed power steering system (standard equipment). It would be nearly impossible to park the car without it. This power steering is responsive to a very light pull on the wheel rim, yet never seems oversensitive or sudden in action. There's no feeling of power "take over" that calls for a tense readiness to back off on the part of the driver.

Power brakes are another standard item and a must to cope with the Lincoln's 5100 lbs. Lining area is a useful 256 inches and the car can be brought to a really fast stop from high speed without much pedal pressure and with no grabbing or swerving. There is as much resistance to fade as the average driver will require.

Under the hood the engine remains much the same, a well-balanced, smooth-running 430 cubic inch V-8. A significant change has been made in this engine for 1960, in that the four-barrel carburetor has been replaced with a simpler two throat unit. This new carburetor makes little difference in the performance of the car in the normal speed ranges (i.e. 0 to 80 mph), although the published gross horsepower output has been lowered from 350 at 4400 rpm to 315 at 4100. Torque output has dropped from a peak of 490 lbs ft at 2800 rpm to a peak of 465 at 2200. This means that the engine delivers as much real pulling power from 40 through 70 mph as it did last year and only the ability to achieve as high a top speed has been shaved.

In return the two-barrel carburetor provides smoother idling, improved economy at realistic cruising speeds and freedom from the excess tuning sensitivity of the previous four-barrel model. Lincoln is to be congratulated for eschewing the advertising glamor of higher horsepower figures in favor of a more practical and serviceable engine.

Lincoln's transmission has, as in the past few years, an optional drive range. In D1 range the car progresses from Low (2.37 to 1 ratio) to Intermediate (1.48 to 1) and finally to high (1 to 1). This is in my mind the best way to operate. D2 starts you off in Intermediate, for a somewhat "softer" takeoff with less chance of wheelspin on slick surfaces, but it is less efficient. The transmission is very smooth in action, with up-and downshifts imperceptible in normal driving.

To sum up, the '60 Lincoln is a car which in the realm of total comfort, ride, silence, lack of vibration, ease of entry, interior room and seating comfort has no superior built on these shores. Perfection has been somewhat overdue in coming to this particular series of Lincolns, but for 1960 it is here in full measure. Really outstanding features are its unsurpassed interior body dimensions and the uncanny ease with which the car handles, giving the impression of a much smaller automobile. In short Lincoln is a thoroughly civilized automobile "that does everything good, like a luxury car should!"

LINCOLN CONTINENTAL

TESTING THE Luxury CARS

FOR SHEER luxury, Detroit provides three magnificent choices, Cadillac, Imperial and Lincoln. No other car in this country and a few from overseas can equal the refinement of these.

Perhaps the best way to characterize the three is in contrast to another car in the same price class, the Chrysler 300-F. It, too, is refined but in an entirely different way. It has been engineered to put the driver in intimate touch with the machine and the machine with the road. By luxury car standards, it has rough, noisy operation.

Cadillac, Imperial and Lincoln are just the reverse. Everything about them isolates their occupants from the road; the whole emphasis is on smoothness and silence. This, as the tests reveal, has its effects on their merits as automobiles.

Each of the three is a different solution to the problem of including a luxury car within a broad corporate line. Cadillac shares its body shell with other General Motors makes but has its own engine and chassis. Imperial and Lincoln follow an opposite policy, using the same mechanical components as their medium-priced running mates with their own distinctive structures. And distinctive they are. Imperial is Chrysler's sole product with a separate body and frame while Lincoln is Ford's only full-sized car with unit construction.

In their regular lines, each offers three series. Cadillac has the 62, Fleetwood 60 Special and Eldorado, Imperial the Custom, Crown and Le Baron and Lincoln the plain Lincoln, Premiere and Continental Mark V. Prices range from $4892 to $7401 for Cadillac, $4922 to $6318 for Imperial and $5252 to $7056 for Lincoln.

All of the closed Cadillacs are hardtops while Imperial and Lincoln each use the same bodies and charge the same prices for four-door sedans and hardtops.

They all build formal cars as well, Cadillac the 75 eight-passenger sedan and limousine costing up to $9748, Imperial the eight-passenger Crown limousine at $15,600 and Lincoln the six-passenger Continental town car and limousine reaching $10,230.

Finally, Cadillac has the Eldorado Brougham at $13,075. Both the Brougham and Imperial limousine have bodies built in Italy so the Lincoln Continental limousine is left with the title of the most expensive car made in this country.

Turning from high finance, the cars actually tested for this report were a Cadillac 62, Imperial Le Baron and Lincoln Continental. Strictly speaking, Cadillac's equivalent to the Le Baron and Continental is the Fleetwood 60 Special but, because the 60 and 62 are the same mechanically, the test results are still comparable.

In every other respect, the three cars were similar. All were four-door hardtops with complete power equipment and air conditioning to equalize any weight advantages. It would have been difficult to have them unlike because so many accessories, power steering, power brakes and automatic transmissions as well as minor niceties like electric clocks, windshield washers, remote control outside mirrors and back-up lights, are all standard. Lincoln goes all the way and even includes radios and heaters at no extra cost.

This may well be a last look at the high-priced field as it has long existed. Just as the Rambler has revolutionized popular car concepts, the Thunderbird has pointed the way to new ideas in luxury. Specifically, it has dispelled the notion that a relatively expensive car must be big.

CADILLAC

For 1960, Cadillac continues the basic design introduced last year with important changes in its suspension and braking systems.

It proved the best performer of the trio, even though it had the smallest engine. From 0-to-60, the 62 test car averaged 10.7 seconds, beating the Imperial and Lincoln by 1.6 and 3.5 seconds, respectively.

The 390-cubic-inch engine is essentially an enlarged version of the first ohv V-8 offered by Cadillac 11 years ago. It develops 325 hp with 10.5-to-1 compression and single four-barrel carbure-

176

CADILLAC interior is least ornate of the tested trio, but basic quality and solidity make it most outstanding within this area.

LINCOLN dash is impressive but confusing home of buttons and dials. Driver has trouble just scanning full range of panel.

IMPERIAL instruments are carried out in same massive proportions as rest of car. Some are more for style than function.

TESTING THE LUXURY CARS

tion. For those who want a bit more spirit, there is a 345-hp modification with triple two-barrel carburetion, standard in the Eldorado and optional in other series.

Behind this particular Cadillac's superior acceleration was lower gearing. With air conditioning, the factory installs a 3.21 rear axle ratio instead of the normal 2.94. This, of course, increases fuel consumption; the 62's 10-to-13 mpg range was poorest of the luxury cars.

At first, the 62 does not feel as lively as it really is because engine noise and vibration levels are so low. An important factor here, and a good example of Cadillac's attention to detail, is the radiator fan fitted to air-conditioned cars. It has a temperature-controlled clutch to reduce fan noise at high speeds.

The coil springs, front and rear, have 10 per cent softer rates this year. As a result, riding qualities are smoother but less stable. The car behaves beautifully on even surfaces at moderate speeds. Across dips and around corners, however, it floats and sways badly.

When travelling fast on a narrow country road, proper control is quite difficult. The car seems to feel, literally, all over the road.

Another annoyance, even on well-paved surfaces, is a vibration that occurs between 60 and 70 mph. Apparently it is caused by a point of resonance in the "X" frame because it disappears beyond 70 mph.

The brakes, too, have been altered. They now adjust themselves, taking up any slack when the car is driven in reverse. At the rear, extended and finned drums improve cooling and allow slightly bigger cylinders. Most interesting of all is a new vacuum release for the parking brake. It operates automatically when the engine is running and the car is placed in gear.

This has several advantages. It makes it impossible for the car to be driven with the brake on, it prevents accidental release when the ignition is off and it provides a true emergency brake. Since the parking brake will not lock with the car in gear, it can be used as a supplementary source of stopping power when the car is in motion.

The 62 four-door hardtop, called a Sedan de Ville in its more lavishly trimmed version, comes in two shapes at the same price. One has a flat roof and four side windows, the other a sloping roof and six side windows. The test car was an example of the latter, which seems more practical because it has more interior height, providing space for a higher seat cushion and greater head room in front.

Generally, the Cadillac does not give the impression of being an unusually roomy car, though its interior dimensions compare favorably with those of the Imperial and Lincoln. One difficulty is its awkward entry and exit, hindered in front by the windshield dog leg and in back by a narrow door opening and deep floor wells.

Interior details are very well planned. The instrument panel is recessed to eliminate reflections and includes the shift quadrant. Minor controls and switches all seem out of the way, yet easy to reach. An interesting touch is the removal of the turn indicator lights from the instrument cluster. There is now an amber light in the chrome strip above each front fender that lets the driver know he is signalling for a turn.

But perhaps Cadillac's most impressive feature is its quality. Not just the excellence of finish or the fact things work but the way they work. Take the power windows, for instance. In most cars, they raise and lower with a whirr, then a clunk. In the Cadillac, they operate with only the slightest hum.

The tremendous popularity of the make (Cadillac sells three times as many cars as Imperial and Lincoln combined) causes

CLEVER DESIGN of the Continental's front doors allows air conditioning tunnel to extend to the rear passenger area. It is the best solution thus far offered to this annoying problem.

177

a low rate of depreciation. This same popularity, however, virtually eliminates the possibility of a discount on the original price, so a Cadillac would not necessarily be the least expensive buy in its class.

A final note: Cadillac is the only one of the three luxury cars available with any mechanical options and even it has very few, the engines and axle ratios mentioned earlier plus air suspension. Other equipment parallels that offered by Imperial and Lincoln and includes an automatic speed device called Cruise Control.

IMPERIAL

Imperial's latest model features extensive restyling of its four-year-old structure but no major changes in its fundamental engineering.

The Imperial engine, most powerful in the luxury class, is the same 350-hp, 413-cubic-inch V-8 used in the Chrysler New Yorker and serves as a basis of the 375-hp unit in the Chrysler 300-F. It has a 10.1-to-1 compression ratio and a single four-barrel carburetor.

The Le Baron recorded a good balance of acceleration and economy for a car of its weight. It averaged 0-to-60 mph runs at 12.3 seconds and fuel consumption between 11 and 14 mpg.

For roadability, the Imperial is at the head of its class. Its suspension system, torsion bars at the front and semi-elliptics at the rear, proves a comfortable ride can be achieved with a definite feel of control. It is smooth on the boulevard yet steady on the highway.

All Chrysler products are noted for their roadability. Throughout the line, their high-speed handling surpasses that of their competitors by a significant margin. However, except for the stiffly sprung 300-F, it is a virtue that lessens as size increases. In other words, an Imperial is not going to keep up with a Plymouth or Dart over a road racing circuit.

But it will stay well ahead of a Cadillac and safely in front of a Lincoln. It has the most stable suspension of any such big, heavy car built today. Float at cruising speeds and sway in corners are both within limits that enable the driver to keep track of the car's placement on the road. If traction weakens in a turn, the classic technique of increased accelerator pressure puts the Imperial right back on course. It responds to correction far more precisely than its two rivals.

This exaggerates a common characteristic of the luxury cars, the ease with which they can be driven faster than intended. Because they operate so quietly, a driver is often startled by the reading on the speedometer. And in the steadier handling Imperial, such a thing is especially true.

All of which makes the Auto Pilot, Imperial's automatic speed device, a most useful accessory. It was the first unit of its kind and, in an improved form for 1960, has become the best.

Basically, it is a governor that can be locked at a pre-selected speed. The original Auto Pilot, together with the present Cadillac Cruise Control and Lincoln Speed Control, had to be re-set manually every time the brakes were applied. The latest version eliminates this slight inconvenience and will re-engage automatically as soon as the car is returned to the desired cruising speed.

Another interesting new feature is an emergency warning light system. With the flip of a switch, all four turn indicators start flashing. The purpose is to make the car visible when it is stopped on dark roads.

Increased emphasis on quality shows up strongly in the 1960 model. For nearly two years, the make has been built in a separate plant where it can be given more careful attention and the policy is beginning to pay off. The new Imperial has the highest quality of any to bear the name in years. It is not yet the equal of Cadillac but, on the basis of the particular cars tested, it seems to have gained an edge over Lincoln.

The four-door hardtop, listed as a Southampton in the Imperial catalog, now shares its shell with the sedan. Aside from the latter's fixed window frames on the doors, there is no real

MOTOR LIFE TEST DATA

1960 CADILLAC

Test Car

Test Car: Cadillac 62
Body Type: Four-door hardtop
Base Price: $5498

Maneuverability Factors

Overall Length: 225 inches
Overall Width: 79.9 inches
Overall Height: 56.2 inches
Wheelbase: 130 inches
Tread, Front/Rear: 61 and 61 inches
Test Weight: 5130 lbs.
Weight Distribution: 54 per cent on front wheels
Steering: 4 turns lock-to-lock
Turning Circle: 47.8 feet curb-to-curb
Ground Clearance: 5.9 inches

Interior Room

Seating Capacity: Six
Front Seat
  Headroom: 34.2 inches
  Width: 61.7 inches
  Legroom: 45.9 inches
Trunk Capacity: 16.4 cubic feet

Engine & Drive Train

Type: ohv V-8
Displacement: 390 cubic inches
Bore & Stroke: 4.0 x 3.875
Compression Ratio: 10.5-to-1
Carburetion: Single four-barrel
Horsepower: 325 @ 4800 rpm
Torque: 430 lbs-ft @ 3100 rpm
Transmission: Four-speed automatic
Rear Axle Ratio: 3.21

Performance

Gas Mileage: 10 to 13 miles per gallon
Acceleration: 0-30 mph in 4.0 seconds, 0-45 mph in 6.6 seconds and 0-60 mph in 10.7 seconds
Speedometer Error: Indicated 30, 45 and 60 mph are actual 28, 42 and 57 mph respectively
Power-Weight Ratio: 15.8 lbs. per horsepower
Horsepower Per Cubic Inch: .83

MOTOR LIFE TEST DATA

1960 IMPERIAL

Test Car

Test Car: Imperial Le Baron
Body Type: Four-door hardtop
Base Price: $6318

Maneuverability Factors

Overall Length: 226.3 inches
Overall Width: 80.5 inches
Overall Height: 56.7 inches
Wheelbase: 129 inches
Tread, Front/Rear: 61.8 and 62.2 inches
Test Weight: 5260 lbs.
Weight Distribution: 54 per cent on front wheels
Steering: 3.5 turns lock-to-lock
Turning Circle: 48.2 feet curb-to-curb
Ground Clearance: 5.6 inches

Interior Room

Seating Capacity: Six
Front Seat
  Headroom: 34.4 inches
  Width: 61.0 inches
  Legroom: 46.9 inches
Trunk Capacity: 17.8 cubic feet

Engine & Drive Train

Type: ohv V-8
Displacement: 413 cubic inches
Bore & Stroke: 4.18 x 3.75
Compression Ratio: 10.1-to-1
Carburetion: Single four-barrel
Horsepower: 350 @ 4600 rpm
Torque: 470 lbs-ft @ 2800 rpm
Transmission: Three-speed automatic
Rear Axle Ratio: 2.93

Performance

Gas Mileage: 11 to 14 miles per gallon
Acceleration: 0-30 mph in 4.3 seconds, 0-45 mph in 7.3 seconds and 0-60 mph in 12.3 seconds
Speedometer Error: Indicated 30, 45 and 60 mph are actual 30, 44 and 57 mph respectively
Power-Weight Ratio: 15.2 lbs. per horsepower
Horsepower Per Cubic Inch: .85

---

difference between the two. Previously, a different four-door, six-window sedan had been available.

A unique feature of all Imperial body styles is curved glass for the side windows. The Le Baron has the further distinction of an unusually small window, creating an effect of privacy for rear seat passengers but restricting visibility for the driver. He has to be extra cautious about things behind him. Fortunately, the fins and taillights are within his view and serve as some guide when reversing the car.

Two circular dash housings contain a complete set of instruments, including generator, temperature and oil pressure gauges. Thanks to non-glare electroluminescent lighting, they are even easier to read at night than during the day.

Placement of some controls seems to have been dictated by styling rather than function. A particularly irritating case in point is the turn indicator switch. It is on the dash, just to the left of the instrument cluster, where it neatly balances the appearance of the heater lever at the right. However, it is quite awkward to reach. A subtle touch of symmetry hardly seems worth making a frequently-used control less accessible.

Other features for 1960 include the square steering wheel also used by Plymouth and the high-backed driver's seat now offered on all full-sized Chrysler products, both items which have been discussed in earlier tests. And Imperial has finally caught up with Cadillac and Lincoln in another detail; power vent windows have been made available as an option.

LINCOLN

Lincoln has carried its current design into a third year with reduced power and new rear springing.

Its engine has the biggest displacement, 430 cubic inches, and now the lowest output, 315 hp, of the luxury cars. A change from four-barrel to two-barrel carburetion accounts for the drop from last year's 350 hp. The compression ratio remains the same at 10-to-1. Variations of the Lincoln engine power the 310-hp Mercury and the 350-hp Thunderbird.

With a 2.89 rear axle ratio, the Continental Mark V test car was good for no better an average 0-to-60 time than 14.2 seconds. This is the slowest figure recorded this year by any V-8, regardless of price or power. Economy was on a par with Imperial's at 11 to 14 mpg, leading one to wonder why Lincoln did not

VENTILATION of the Lincoln is greatly improved by the roll-down rear window. Driver controls the power operation from his seat. The glass slides into the area behind the rear seat.

challenge the big Chrysler product in the Mobil Economy Run. It was the only make not entered in the 1960 event yet seems capable of a very good showing.

The rear suspension of the 1960 Lincoln is completely new. Previously, coil springs were used but these have been replaced by conventional semi-elliptics. The coil system had been designed so that air suspension could be installed with a minimum of difficulty. However, the option is no longer available.

With the same spring rates, the ride is just as smooth and much steadier. The car does not squat as badly as it did under hard acceleration nor does it dive as sharply when the brakes are applied. There is still some float at high speed but considerably less than there used to be.

In corners, the new Lincoln handles better, though it is still no match for the Imperial. Tires squealing and body leaning, it tends to plow through turns because of its nose heaviness. Approximately 57 per cent of the weight is on the front end.

A redeeming factor, though, is the steering. It is one of the best power-assisted units available. It has enough road feel to please the skilled driver and is remarkably quick. In normal driving, it is even faster than its 3.2 turns lock-to-lock might indicate because the car has a smaller turning circle than either Cadillac or Imperial.

Under traffic conditions, the Linclon is surprisingly easy to maneuver for a car of its size. Crisp fender lines make it less difficult to place in crowded situations than its competitors.

All this is despite its immense bulk. It is the longest, heaviest (and costliest) six-passenger sedan built in this country. It is within an inch of being 19 feet long and has a shipping weight over 5,000 lbs. As tested, it was the heavyweight champion of the year at 5300 lbs.

Interior dimensions are consistent with its overall size. Lincoln has higher seat cushions and more head room, front and rear, than either Cadillac or Imperial. Much of this can be attributed to its unit construction, which allows a greater distance between ceiling and floor because no thick frame members take up space under the passenger's feet.

When the body and frame are combined, a car can become more subject to noise and vibration from road shock. Lincoln has avoided this problem with thorough insulation. Even on the roughest surfaces, it is a quiet car. The Continental tested did have some wind noise at high speeds but this was due simply to an ineffective window seal.

Lincoln calls its four-door hardtop a Landau and, like Imperial, uses the same shell for its sedan.

The Continental series has an exclusive roll-down rear window controlled from a console in the driver's arm rest. The glass slides immediately behind the rear seat so there is no room for a rear radio speaker but it provides wonderful ventilation. On all but the hottest days, it is more pleasant to leave the air conditioning off and lower all the windows to enjoy the forced breeze that comes through the car.

Turning to the dash, instruments and controls are gathered together on a single, flat panel in front of the driver. Heating, ventilating, defrosting and air conditioning are all combined in one of the simplest, easiest arrangements possible. A single knob controls fan level and a dial setting for the kind and temperature of air wanted.

The parking brake has a vacuum release, similar to Cadillac's, that will not function unless the engine is running. However, it is operated by a pushbutton rather than by placing the shift lever in gear.

Two small features will appeal to the family man. The electric door switch is on the dash, right below a red light that goes on if any door is unlocked. Thus, if a child unlocks a rear door, which he must do before he can open it, the driver is immediately warned and can relock it with the dash switch. And the master control for the power windows has a control that cancels operation of all but the windows in the driver's door, preventing small hands from playing with them. ●

MOTOR LIFE TEST DATA

1960 LINCOLN

Test Car
Test Car: Lincoln Continental
Body Type: Four-door hardtop
Base Price: $6845

Maneuverability Factors
Overall Length: 227.2 inches
Overall Width: 80.3 inches
Overall Height: 56.7 inches
Wheelbase: 131 inches
Tread, Front/Rear: 61 and 61 inches
Test Weight: 5300 lbs.
Weight Distribution: 57 per cent on front wheels
Steering: 3.2 turns lock-to-lock
Turning Circle: 45 feet curb-to-curb
Ground Clearance: 6 inches

Interior Room
Seating Capacity: Six
Front Seat
  Headroom: 34.9 inches
  Width: 60.4 inches
  Legroom: 44.0 inches
Trunk Capacity: 17.2 cubic feet

Engine & Drive Train
Type: ohv V-8
Displacement: 430 cubic inches
Bore & Stroke: 4.3 x 3.7
Compression Ratio: 10-to-1
Carburetion: Single two-barrel
Horsepower: 315 @ 4100 rpm
Torque: 465 lbs-ft @ 2200 rpm
Transmission: Three-speed automatic
Rear Axle Ratio: 2.89

Performance
Gas Mileage: 11 to 14 miles per gallon
Acceleration: 0-30 mph in 4.9 seconds, 0-45 mph in 8.3 seconds and 0-60 mph in 14.2 seconds
Speedometer Error: Indicated 30, 45 and 60 mph are actual 28, 42.5 and 54.5 mph respectively
Power-Weight Ratio: 16.8 lbs. per horsepower
Horsepower Per Cubic Inch: .68